The Forensic Economist's Guide to Cryptocurrency

Charles W. Evans

Pecuniology Press

First Printing, 2019

ISBN 978-1733-6036-07

Dedicated to all who seek
truth and justice.

Table of Contents

Foreword

As the title implies, this is a guide to cryptocurrency for forensic economists. It is based on material that I have submitted as an expert witness in criminal and civil cases involving Bitcoin and other cryptocurrencies in several US state courts, Federal court, and US Army Trial Judiciary. It is a distillation of the topic to the salient issues that arise with reasonable consistency in a broad variety of cases, along with some economic, financial and high-level technical background.

Although thousands of different cryptocurrencies circulate, this guide focuses primarily on Bitcoin. It follows the convention of referring to the software application and the underlying protocol as *Bitcoin*, written with an upper-case **B**, and the units of divisible electronic 'tokens' as *bitcoins*, written with a lower-case **b**. Nonetheless, the discussion herein generally applies to all of the most popular cryptocurrencies.

If after reading this guide, you have further questions, would like permission to reproduce it in whole or in part, or would like help preparing a forensic economic report that involves cryptocurrency, do not hesitate to contact me by email at: cwe@EvansEconomics.com.

Charles W. Evans, PhD (Finance) MA (Economics)
Managing Partner, Chyden LLC
Ft. Lauderdale, Florida, USA
January 2019

Introduction

Since its inception in January 2009, Bitcoin—and the subsequent category of *cryptocurrency* that it has spawned—has grown from a computer science experiment with no market value into an industry worth approximately $150 billion over the following decade[1]. In that time, civil and criminal cases involving Bitcoin specifically and cryptocurrency generally have been tried in the USA and abroad, and it is not unreasonable to expect that the trend for this will be to continue growing along with mainstream adoption of cryptocurrency as a medium of exchange and a store of value. Given its relative novelty, lawyers, judges, and juries rely on expert witnesses to explain Bitcoin to them from the perspective of accounting, economics, and finance.

This guide is intended to help forensic economists familiarize themselves with the salient aspects of Bitcoin specifically and cryptocurrencies generally. The target audience is practicing forensic economists, who might be called on to serve as expert witnesses in civil or criminal cases involving Bitcoin or one of the other cryptocurrencies.

Although, cryptocurrency is increasingly important as a medium of exchange and a store of value, none is used yet as a unit of account or a measure of value, due to the volatility of their market values relative to conventional monetary assets like gold and silver. Civil and criminal cases involving cryptocurrency typically involve accusations of money laundering, unlicensed money transmittal, fraud and scams, conversion, hacking and blackmail, fraudulent conveyance and non-disclosure of assets, providing material support to terrorists, evasion of economic sanctions, etc.

The presentation in this guide focuses on economic and financial matters, leaving aside technical matters that are

---

1 One must be very careful when referring to these numbers, as they rise and fall frequently over the short run and tend to move generally upward over the long run.

largely irrelevant to the forensic economist's testimony. However, some high-level technical background is provided here, in order to demonstrate how Bitcoin is innovative, and why it is commercially interesting. The material presented here has been the basis of pre-trial interviews, depositions, and sworn testimony in several US State courts, US Federal court, and in US Army Trial Judiciary.

Readers unfamiliar with this topic should find helpful background information at Bitcoin.com and Bitcoin Wiki[2], and in Brito & Castillo (2016), Grinberg (2011), Narayanan et al. (2016), and Velde (2013), all of which are listed with links in the References section at the end of this guide.

---

2   https://en.bitcoin.it/wiki/

1. Cryptocurrency

The term *cryptocurrency* is a portmanteau of *crypto-* from *cryptography*—the science and practice of secret codes—and *currency*, referring to money and money substitutes. It, and terms for its attendant concepts, are metaphors used to refer to a category of software applications and protocols that manage what can be thought of, metaphorically, as commodity *information tokens*. These tokens can exist for no specific purpose, as was the case with Bitcoin, when it was released in January 2009; they can carry a promise of some value, as when they represent, e.g., pre-sales of goods or services, transferable points in a game, or shares of equity (i.e., they can be used as derivatives that are backed by some underlying asset); they can be thought of as units of information in a messaging system, as they might be employed within a bank or a network of vendors. The range of possible uses and concomitant metaphors is limited only by the imaginations of those developing, using, or analyzing cryptocurrencies.

From the perspective of the software application and the protocol—as well as from the perspective of accounting, economic, and financial analysis—it is largely irrelevant what term one uses to refer to a specific use for a cryptocurrency. It is what it is, it does what it does, and it is worth what it is worth, no matter what one calls it. However, one's choice of metaphors can have profound statutory and regulatory implications, depending on whether one chooses to describe it as, e.g., money, property, commodity, pre-payment, electronic tiddly-winks, etc. (See §2.B. "Bitcoin in Practice" below for a more detailed discussion of this point.)

When one uses terms such as *cryptocurrency*, *electronic cash*, *virtual currency*, etc. to describe Bitcoin and similar software applications and protocols, one most commonly is framing it as *money* in the broadest conventional sense; typically as a medium of exchange and a store of value, and generally less so as a unit of account and measure of value. A popular metaphor among Bitcoin users is to see it as 'digital gold', in the sense of a new class of ephemeral *asset*, *commodity*, or *property* that can be used in barter *as money*. Note that this interpretation has been accepted by state, federal, and US Army courts, in the absence of statutes, regulations, or administrative rules to the contrary. The forensic economist testifying as an expert on cryptocurrency should double-check the legal status of the jurisdiction, in which he is testifying. Things can change.

1.A. The History of Cryptocurrency

The roots of cryptocurrency extend back to the 1970s, with the development of public key cryptography, which enables one to provide a means for anyone to encrypt a message that only the recipient can decrypt. Previously, both the sender and the receiver of the message had to hold copies of the same encryption/decryption key that they both had to protect as a closely guarded secret. If either party were subverted, then the communication channel would be no longer secure. With public key cryptography, only the recipient needs to protect the secret key. Holders of copies of the public key can encrypt messages that only the recipient can decrypt, and it is common practice for the recipient to publish the public key.

Similarly, the holder of the secret key can use it to encrypt a message such that all holders of copies of the public key can decrypt it, thereby confirming that only the holder of the secret key could have encrypted the message. The holder of the private key can append such a block of 'ciphertext' to the message in clear text, using it to serve as a signature that certifies the identity of the sender of the message.

Combining public key cryptography with hash functions that create unique strings of characters that serve as 'fingerprints' of arbitrarily long strings of information, software engineers developed the first 'electronic currency' or 'e-currency' systems in the 1980s. They began making these systems available to the public in the early 1990s, following the invention of the World Wide Web.

These first-generation e-currency systems ran on centralized servers that were controlled by trusted parties. This made the system vulnerable to attacks against the operator, and put users of the system at risk of the operator's potential incompetence, negligence, malice, or corruption. Nonetheless, centralization was understood to be necessary at the time, because software engineers had proven with logical and mathematical rigor over the quarter-century from the mid-1970s to the early 2000s that a decentralized computer network could be subverted by as few as one-third of the computers on the network. Even if the system employed checks and balances among trusted parties, a minority of them could hijack the system and counterfeit transactions.

In November of 2008, an individual or a team, under the pseudonym 'Satoshi Nakamoto' (roughly, the Japanese equivalent of 'Joe Smith'), released a white paper on an obscure discussion board that described an electronic token system to be operated on a *distributed network* of computers[3], in which no party need be trusted, or whose identity even need be known. Satoshi Nakamoto released the first version of the software that implemented this system as free and open-source software, and it began running on the first users' computers beginning 3 January 2009. It has been running as a worldwide distributed network on users' computers ever since.

The innovation that obviated the received wisdom concerning the viability of a distributed network of untrusted parties—with no CEO, no headquarters, and no authority other than the algorithms encoded into the software running on users' computers—involves 1) a means to ensure that only one party at a time can broadcast transaction confirmations to the network and 2) the distribution of the complete transaction history ('blockchain') to all network participants. By providing copies of the transaction history to all network participants, the system enables all of them to confirm that their copies are accurate and to identify any rogue participants who try to counterfeit trans-

---

3   See §1.B. The Byzantine Generals Problem.

actions. The downside is that the Bitcoin blockchain contains nearly 200 gigabytes of data as of the date that this guide was written, and it grows by approximately 120 megabytes per day.

The deep technical details of how a cryptocurrency system like Bitcoin works are beyond the scope of this guide. Readers who are interested in understanding better what is going on under the hood should find Antonopoulos (2015), Barber, Boyen, Shi & Uzun (2012), Kroll, Davey & Felton (2013), and Nakamoto (2008) to be helpful. Also, as Graef (2010) points out, the reader should be wary of the neologisms used by cryptocurrency users. In the same way that early automobiles were called *horseless carriages*, electrical systems were originally described using images from plumbing, and the Internet was once called *the Information Superhighway*, cryptocurrencies are not *currencies*, as understood by monetary economists; *miners* (discussed below) do not mine 'ore' that is 'smelted' into 'bullion' that is 'minted' into 'coins'; and the *coins* in Bitcoin do not come in fixed denominations.

Even though the forensic economist need not be an expert in the mathematics of cryptography, protocol design, and the finer points of computer science, a high-level understanding of the underlying technology helps to establish one as qualified to testify, under oath, on the topic. For our purposes here, one can argue that an understanding of four salient topics helps to distinguish blockchain-based cryptocurrency systems like Bitcoin from other payment systems: a) the history of digital money or e-currency systems, b) the Byzantine Generals Problem[4], c) blockchains, and d) ephemerality. Each of these is discussed in turn below.

---

4   See §1.B. The Byzantine Generals Problem.

1.A.i. Technical Prehistory of Bitcoin

This section addresses technical issues mentioned above. It goes beyond what a forensic economist should be expected to testify on. Nonetheless, familiarity with this material is preferable to answering, "I don't know," during conferences with clients, and especially under cross-examination.

1.A.i.a. Public Key Cryptography

As mentioned above, Bitcoin's roots extend back to the 1970s, when Diffie & Hellman (1976) first developed the mathematics underlying public key cryptography, and Rivest, Shamir & Adelman (1978) released the first public key cryptography software, known as RSA (the developers' initials). With a public key cryptography system, one has two strings of random characters (keys), such that one can use one key, known as the public key (see Figure 1) to encode a block of legible text, known as cleartext (see Figure 2), rendering it illegible, and use the other key to decrypt the resulting ciphertext (see Figure 3) into the original block of cleartext.

One can share one's public key freely, however one should take a similar degree of care to secure the corresponding private key that one uses to secure the keys to one's home or to a safe. The keys are symmetrical, meaning that if one uses one key to encrypt (encode) cleartext, then one can decrypt (decode) the resulting ciphertext with the other key. In other words, calling one key *public* and the other one *private* is a convention, and not a mathematical necessity. In this way, anyone can use one's public key to encrypt a message, and only the possessor of the corresponding private key can decrypt it. If the ciphertext is intercepted *en route*, it is meaningless to the interceptor (see again Figure 3).

Figure 1
Cryptographic Public Key

-----BEGIN PGP PUBLIC KEY BLOCK-----
Version: GnuPG v2

mQINBFZDmzoBEADYpqLayAEq6UACkht/CNwHqvAVVhJx+RMq6mUlHK00CR+MrO1q
sUsWCYyWmXjObmEsbWjjPprmapfF5UJqueTNPhDr9sTWQjUTWmZmSTKoO+fnsTEO
bY98uiV2HIwMOLmEBoE8imfcOoBJHLQOaKMiq/oAprw+bO69/cz08oRrMbeoUxw9
o66gNTWILhZ0byv+ymOoadhy9jRJL3rY06w8QkMUA/JNmGfpcPNq+llxoaycbfAr
Tq/crKOf8wH5uK7PuF7xUfCaw8s+Lq2kmgDodG2hyT8RmFQjVe6leP+xxPj9Pria
vUXZoNnYf6YWneZVtgXt5z5eD2HcT5hYwO7oxlvHTr5VazAC75B3pJnQ8em1G4F0
Q3ImqiHFu4fN+wJ0dWj4Q33NQF9GuRy133uozgjpn7HaUyFsswxL3Ag3Bh34aZIG
53HBoUkzl44v4EisuuMBfc5ZXXBbLy3twgCwyvI20Z1xjR5sMqU9ThrfcvwMnRiy
PmHqiTNF240MTer0LgtPV07cAgu7bWzDYba6cdnPSMl86nrOV7E8VX6lws2bPLE9
4K1q1sZV4CbFnz7sgMLMJrJB9TUtpBDkSP10PcXl2vvC17EmoEp/FD3Z+eeWJZ21
mnLTDnEY9fwLgFvciLd3tNurCKl8uJVop/AlfWG775fEwxRi+p2yn9vIFwARAQAB
tDhEci4gQ2hhcmxcyBFdmFucyA8Q0V2YW5zQENvbnNiaW5zc/k1c0VudHJpdHylbmV1
cnNoaXJzLmNvbT6JAj4EEwECACgFAlZDmzoCGwMFCQHhM4AGCwkIBwMCBhUIAgkK
CwQWAgMBAh4BAheAAAoJEL5W76F2L7q+VHAP/iVs0Cp8VtlQuESsg9qoflqB/+YiJGMX0y+t
pAEnhqdRYf/8rQpazTCTToOHrP2+76ZE2MzaMBHWxZ3DLuHhfgWkc7VnJUT8w6CP
OMahnMuqI8Crur/cIEVQMJrfAHrpcC6FHKM7ukGbkinUDCH5VaNnzir6iJ11bC4H
uYggY+dM7iSeOJfoRJJN/0OBYQckTkE3ZwEPWhANZOSNvWz5G5pgGN810n/Nc6Ox
O27u/lEjkqx6IxSceuwBuqSOIRyxLpMtE9Qw3fcPriQbCkadILLiQeF0Z9YQVD05
4HMuP3zDFh8ygFx6te1vg1ZN/0uZE39lRa7klp9/heLTbvMVyKHbncildovXqzba
gb8t7y8DdwmF7MRgDlaxtb62ZRLmLIU5S+6DdfbOChO0wdjELpCF3/ZSwIB4QFs0
GxRrcR0hElwRn8rmqbLLxTgrKpp61//juwdQgl96JCGxKR+UIRMhDKMneGmO5JYF
ct7iQQ7MJYbqnzsIjWTM+2rkt/1/8uy1cocU/6eVOo+mKWIeQkSgAhE51VeLgD1B
5UXHJYWgGz9XbpGtMELzpcO/wO+bsSKwX5EW2vqf+nw5nfBejzLmILCmf8JCKkBW
61EZHOttdO0exzCTuz2a6mLSTt0Mppe5Hw1R+MdqaQ6vJg666vTu8bq0dDjiaBj4
mq5wvh4ttCFDaGFybGVzlEV2YW5zlDxjZXZhbnNNAY2h5ZGVVuLm5ldD6JAjwEEwEI
ACYCGyMHCwkIBwMCAQYVCAIJCgsEFgIDAQIeAQIXgAUCVkiwkAIZAQAKCRC+Vu+h
di+6vqlxD/9v6yALeFy+MXX/PBeMbjTjKSNTr2nmE/Fv7y4ZlZM85S6ldZ88TTSN
LlnAEGcCnIbS4e5MRv+1J/fcXF9I5Pl3xtHKikfiG1nutqcIO+K+laCETXpHu0jI
VEHiPaeWRTrJKk+LT/+5OTcdMgPfmeVgdOZgWVC0GkjNOFWLdxDfbOgKdNCkz+qe
pB7J9QrXI0Cyo9V1eICzES/8kOXuz/oZgKJ0DXUZ6sAQ1ad62aLv5IfqkqmBRf5I
DJfkIqQxBt/xrRX88OivW8sxd0sdiu7u2oyYSBVYGKrjqmHmvAXkOpDERowx91AN
TK4H6V3WSMMi1ctCP3UyLgthzHxau7ILlXt7TeD9P6ibS6P7MSwR3boNDdNyfi7x
Og6SnnNoOS3C2IXR3YrdToKRaAYNJlCEG3oRLH8IcxDm+mWIbQM/4j5lge0xZbOB
hBK5XAgunqDTob/1oXSbwQBVWcQJuv7RdOLlnF76gKneH0vlrntxNy8NAsVYSUNd
bSZAgzGeGl06XhaYw17TL18XnYEgACfx8x7M5XhByGiIIVyM5yi7bLLUkyjmq5Hk
awlXHNpooUav/xCuDtRIZOEdYln3GUjxPrvoPWikKSSlQt/5Q/TbS+43hSjxbJv7
B2h8hWihS8eDpy8zSkdQNfWGia4LT0T+T3tRwnwoWiRuU6k3XdOaLYkCOQQTAQgA
IwUCVkObOglbIwcLCQgHAwIBBhUIAgkKCwQWAgMBAh4BAheAAAoJEL5W76F2L7q+
u5UQAI3hV7871soPGFLpc1VqtKGCQGJvdL8XUgEGpLtb9NbXxWkL6i9606+fewcf
0lGZDCOLTa7CNgPTO4PK7tomoXVPhsldB1OYeLPvKo1LXMcRfsDMlHZWoG684YQL
ISoQqudb3OlizMlRGtrkEyOUD8+y1nQu3MZW0204c7W4Qd0HDKxn868ldmJFAoWF
JA3XMWwtt0lYCdYu5517fBXZrQJriENdlIQOs8mwAd4HeXAQbkx9LuGp39b+mrv/
qG2f4q4cpTF3p9SgU75N64XhB7EHmSSrR08xAWdL2e7OfawBTy7P8Ym9l9xnqDrb
SXeKsqDepuSB2S+sXVWfa4rgR9GuyUB2u4jot4KhI7C7CV//bqyB21Bj9g88pNEr
jn6KerrhgmL6gxVWu3W7dvdVNKs5yAZFzHAstNBh73CoLy9Y16/W1b8TcRHB1RYL
rN5a5TuFNWzljC3IOrZxd5hilU8YiHhATmVuiDxoBVLA7KomUPVEunyH+q4RAeG6
2NLqkhpPgScyc0ugE1p6NsZPlNmjTUG3q+ReqPVck2fzeIr1NYpcDA0rVAb3woeE
4qzueGa1HKu7gAv3IIUY0xJo+ssbiBN2i7090KADQUwYrsNxsbV4f2ouJ90IKHMK
CkTSc7C4hIN0amH1lO8WBjqrt+MnKyLBzB8F+2FpZe3q8s81uQINBFZDmzoBEADF
HRflK6Jo1FCZLk7AOQf+tgcbWJy7ATOw51DKXT+gRU57nEajuN0nhsteg/z06Dnm

6

m6jZGdcBHauPVPMO2pIaOgTYSgasRfx/DvyXqdDyr1AUu2iSdVz6meFbZnxMdTY5
JDhxCPC4KQGD0TjhaXCng55PBBFtDGji+0lK2Xier2TYwtxq7DfeVwr5XQTIgwqf
4egoz0J833Bv6M9wgksR0bFqzFULY5fVN+opAGxfD8oJSUtyPSNqKGGXUEnAfvwD
7cKF2+Wt1ziewnSjTCluR8uLRTuPaqYwvaSa07XU7bAZywgB5y1GtAoywKV3V7ql
z05LZcceSZFHajaUfiFv5rxDMYlRGixY/XkgFuxq9cuzZ9x/hOslSHHADT+qnGhj
uUclLXjLtcQ87/dG16hBnEMJNi7yCVanmr9UH6MJEsXZrFmiRVIe3NrejZZIKInj
kZKCSMAh8S/H5wZ3S2V/OSz8mWh/NJ2quXq0QX5JPcJzdPABMgWUnJjGx/63MG3v
738LAB5h+m6x4/cds5NDDflV39cWroYV+eV6kOe3ZJkumDZ8pFKnUfaUozvJ6DhQ
CjWD4WOjnyAoRZ8pTkBss12Pyi5hyAZsTqgHrKp6P2OXuxgnAaG2ipX0r5E1LETq
OKjvv9FKYpXIk4v7Mwpe7mo6A5tnru4W0shQuj5XqQARAQABiQIfBBgBCAAJBQJW
Q5s6AhsMAAoJEL5W76F2L7q+/0YQAKaI81SJBYtNdzkP5F3XHxOOtREzVZ74rN6P
VdsOU/qeOx3DF4iiAayU82GRPTP/Q/bQpeqCz9Aa8JSXksupgFJN6+CYErMBH2Ct
f7boCPmkHaPRuB7b2te1UOF9wmihW4Vu80LO7C923CwNmJRmax5GTQ+J5cvtTNur
hQ0VOHwdi7N0VbP8Gke0G370vnygFfhQFI6OgVSqETf8RUP7u1Qe+bcvWMjxUPmU
g7VzpQvoDcbqV+Zktnlrt2x8PZcG/7+fvjL0it+VP/H073MzEgadzH7tJNrhjJEj
9VVzMSgWyEoJlyAuk04gAH7bVs8QbRcj3vnry7aBfNRzEEjx8lFCYwfLe/88NEoL
+zqg1SRO/gowPQLDYdreqHTMqBFvKTqJc0lbGGUJC8740+HGIG9IafhwWRX9pird
mqFuhagiSN0mgAdvcP50pdzRfd6lvUZJACXrGT1Yw9l2/7oCt42OMW9iVfaL2F4m
BN9ayK99iC2mpNClZIUSFD/AT4s9p9gDi1qVW+nu8iXUAF76YjcvDnwxWjzhn/j8
l8UBezhSQK2ojf46CkyS5Tun9ndMbLhpuK7bMTmF0gAGsJXpQOJquW0ugcrc4Tch
HD3GgvxzVjtR/Bl7+7M1wLFHLMUfmwLeE2SXXzZS+KYLnEKn8GhE58scwrWeT52r
7GsXx37g
=zfao
-----END PGP PUBLIC KEY BLOCK-----

| Figure 2 |
| Cleartext |
| This is an example of cleartext. |

Figure 3
Ciphertext

-----BEGIN PGP MESSAGE-----
Charset: utf-8
Version: GnuPG v2

hQIMA6/DreKSJ9cXARAAttRQ84cCdSa17D5ft6aujGhx4CCT2FWY1B48MYbKV0F+
R7BItdkHPJIIVLttsNUo2JfpsAFzwxTv4qFtzs4LL+Fou/ighHRHTUTLFWZ4peKF
BNU0TqcrtPc3j1vDTubQXqKTNV21dVz2x07ZiAGjeFM0ZDVXANcHlntBhpu+K+KX
jwbJS271UegHAZEqatD3ue9HezyWn3BgvoTnh9ZoCwcEsuI+rbqVSusmPdvrtKBF
8ajpXtPtQzmQFIDp4nTaR+orQvIm7aeodgRphfB+dC0DMO2KlQac2jyJ7UmMwTV7
zKUoDG7M1oFoLILjUb7Ip5+dfeGehoVeBIsAr2IZXZnBLC6Dh87PX8BA87fKeLHZ
9QK3ikFvsiNEzD3CCR0Q1cm20BZt2shjWCQCa0fqWhh+5IXBMQ3hz6gBcYWt4nRR
rll5p6hY/8woaewxkO45lCTo5ZFpndt+lpYJmozL4wdQytOCXGTP6tEZhl0Blj7W
vU58uKMAteGFwAiyqImlkBIPPA6HPB5VOiKIaUt5XGr7cipkTXaKB0naeloDTpLO
Ai6pnwyvCCujodhdWxqQ3mi64VuZ4HNDhJLWLR1Hn6HcWmYf/ygqbWNuLwdg3ETd
fedqXqXHWy0SIg/bV1LWAMzWno2KP/YsncPZkSKt5NqucgGSTerd3ZoYxgU67pLS
6QFeCa99kw4UhzsQh7IKsuEtC2LGL0p7B6YY/icTKwC6xpWFfJ1gk0xogB+wy0rl
2jo+p/uDFlGHR3EUdjjkrDkZ5vxqjY13aPUSYmByx1BGPhjVp8NIIF90M9suWyQv
908WVJqXOPxjeJ4/MGRVirkv2DrXbmZKMdPnvldSX67sTBOkd7TUMLBblqSuZogv
mvlARtw52GqaOtP0lvnqCbcjh10dOFLvT0DD9BfLZa6+pTC1siRu+Adft4g1KHjy
nubPZuRurDXeOtM7WJ2zlyTsO60VtPW20/sJYpOTWsF7evggZFuogn+/A+bpur6T
pq4Vm+k+KC8rGtME/ZIyhroKeAFTiL908Rv9Et4sWwXP8P/IFDT4dvJGpKE4g6AE
+m2R8l1NFWULzZlfp6YUCz0HCnk5Ys7hUljFBXoJz8k6kvMXVapRqRu0PadPTpWl
YgspyD4dgOV3CWHQyVA8Z4Gd5+mxaQuC3QGzuaDOMN+u8V5baPvpXkfYHWgqc4F1
FycrcqYCCA/hpqA3lEPwgl6zO9KGf4R/4gS/aHidg9YX3BUTqZQuYcMbWEPhyEyy
KApsJerqvX0RzdzNi7gttuae4j0trsm8vPoEU05qk4ApM7gxrDIR5KYcFvacDKAk
O0DjgWlKoJVFoleJdXCtS9F1+NZPT8fAjJUgKxyZS2jqjhKSE5WWq7EHzrLM0cd8
a0oEz2ODQSrZTFTcsW7E3j05l8lRba8N3ryMEEVGbWFegErV0f8nIS/xIKfCW6l/
OOdtuOcHw9OePjGHp4eeq1mJLl7ByS1KkuRlEHixztOsdHCbalb8Z8C6sBjpKWVJ
Sy4yM7tiC+gitxLLfmkeAC4C6OAivUM7lMEcWUQevdg=
=+7qj
-----END PGP MESSAGE-----

8

1.A.i.b. Cryptographic Signatures

One can do things the other way around and encrypt a cleartext message with one's private key, thereby creating a *cryptographic signature* that one can append to that message as proof of authorship. Anyone who has a copy of the corresponding public key can decrypt the ciphertext and confirm that it is identical with the cleartext message and that only someone who possesses the corresponding private key could have created that cryptographic signature (see Figure 4).

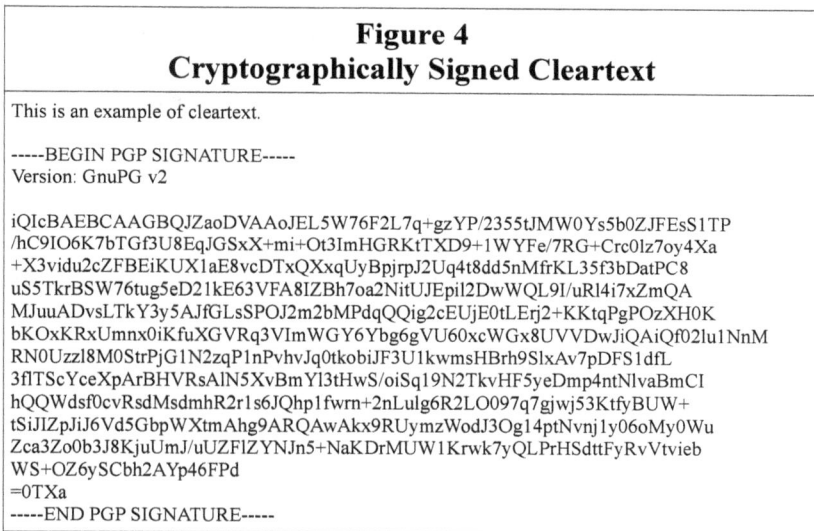

Figure 4
Cryptographically Signed Cleartext

This is an example of cleartext.

-----BEGIN PGP SIGNATURE-----
Version: GnuPG v2

iQIcBAEBCAAGBQJZaoDVAAoJEL5W76F2L7q+gzYP/2355tJMW0Ys5b0ZJFEsS1TP
/hC9IO6K7bTGf3U8EqJGSxX+mi+Ot3ImHGRKtTXD9+1WYFe/7RG+Crc0lz7oy4Xa
+X3vidu2cZFBEiKUX1aE8vcDTxQXxqUyBpjrpJ2Uq4t8dd5nMfrKL35f3bDatPC8
uS5TkrBSW76tug5eD21kE63VFA8IZBh7oa2NitUJEpil2DwWQL9I/uRl4i7xZmQA
MJuuADvsLTkY3y5AJfGLsSPOJ2m2bMPdqQQig2cEUjE0tLErj2+KKtqPgPOzXH0K
bKOxKRxUmnx0iKfuXGVRq3VImWGY6Ybg6gVU60xcWGx8UVVDwJiQAiQf02lu1NnM
RN0Uzzl8M0StrPjG1N2zqP1nPvhvJq0tkobiJF3U1kwmsHBrh9SlxAv7pDFS1dfL
3flTScYceXpArBHVRsAlN5XvBmYl3tHwS/oiSq19N2TkvHF5yeDmp4ntNlvaBmCI
hQQWdsf0cvRsdMsdmhR2r1s6JQhp1fwrn+2nLulg6R2LO097q7gjwj53KtfyBUW+
tSiJlZpJiJ6Vd5GbpWXtmAhg9ARQAwAkx9RUymzWodJ3Og14ptNvnj1y06oMy0Wu
Zca3Zo0b3J8KjuUmJ/uUZFlZYNJn5+NaKDrMUW1Krwk7yQLPrHSdttFyRvVtvieb
WS+OZ6ySCbh2AYp46FPd
=0TXa
-----END PGP SIGNATURE-----

Public key cryptography underlies information security in the banking system, in military and intelligence organizations, and throughout industry and government. The most common software for the management of public and private keys is called GPG. It is an open source version of Pretty Good Privacy (PGP)[5], which is a commercial software product. GPG is

---

5   Cryptographer Phil Zimmermann created PGP and released it to the public in 1991, in protest of export restrictions at the time that classified cryptographic software as a munition under the US Arms Export Control Act, but allowed source code to be exported in printed form.

freely available for essentially all computing platforms, and it is incorporated in many email systems that encrypt and decrypt messages automatically.

1.A.i.c. Hash Functions

A related topic is one-way hash functions that operate on blocks of cleartext of any arbitrary length and return fixed-length strings of random characters known as *hash values* or simply *hashes* (Schneier 1996, 359-383). These also are known as cryptographic *fingerprints*, because each block of cleartext has a unique hash, and the likelihood that any two legible blocks of cleartext will generate the same hash is so negligible that is effectively 0% in practice (see Figure 5 for examples of hashes). Hashes are known as one-way functions, because it is very easy to calculate the hash of a block of cleartext, whether it is an instruction to a subordinate, a contract, or any other cleartext—in fact, one can find hash calculators online with a simple Internet search—but effectively impossible to generate the original cleartext from the fingerprint.

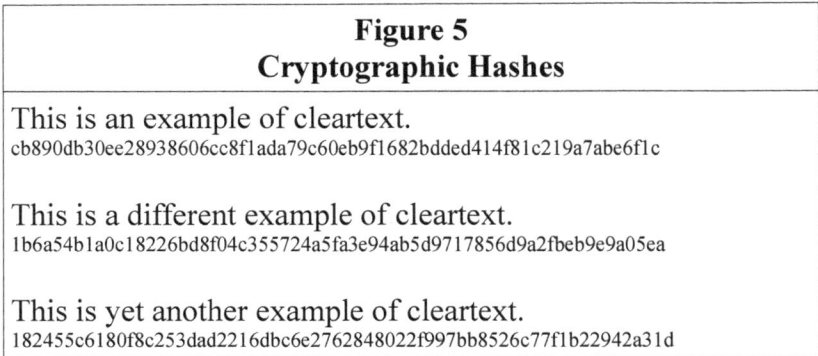

Figure 5 **Cryptographic Hashes**
This is an example of cleartext. cb890db30ee28938606cc8f1ada79c60eb9f1682bdded414f81c219a7abe6f1c
This is a different example of cleartext. 1b6a54b1a0c18226bd8f04c355724a5fa3e94ab5d9717856d9a2fbeb9e9a05ea
This is yet another example of cleartext. 182455c6180f8c253dad2216dbc6e2762848022f997bb8526c77f1b22942a31d

Used together, public key cryptography and cryptographic fingerprints can be used to construct very powerful information systems. For example, one might draft a contract, have both parties cryptographically sign it, append the two signatures, and then calculate the hash of the block of text that includes the contract in cleartext and the ciphertext signatures, and then publish the fingerprint, e.g., in a newspaper or other reliable outlet with a verifiable publication date. While no one will be able to decode the fingerprint, if a contract dispute arose, then either party would be able to produce the original contract and signatures for a judge, arbitrator, or mediator, and show that the fingerprint existed on or before the publication date, proving that the contract and the signatures existed on or before the fingerprint's publication date.

1.A.ii. eCash

In the 1980s, when the Internet was still the purview of university students, faculty members, and military officials, and essentially unknown among mainstream consumers, David Chaum (1988; 1985; 1982)[6] described an untraceable electronic cash system that built on the work of Diffie & Hellman and Rivest, *et al.* This was based on blinded cryptographic signatures that enable one to sign a message digitally without having access to its contents, such that the signature can be verified by third parties who have access to the message's contents. Later, on the eve of the Internet's becoming accessible to mainstream consumers, Chaum, Fiat & Naor (1990) addressed the issue of fraudulent double-spending, which had been one of the most difficult issues for computer scientists and cryptographers to resolve, and that held back the release of a consumer-grade 'e-currency' system.

This research led to Chaum's establishing a company called DigiCash that released a commercial software product based on his theoretical work, called eCash. At the time, the idea of transferring money securely over the open Internet was revolutionary, but DigiCash ultimately went bankrupt in 1998 apparently due more to managerial missteps than with technical or financial shortcomings (Levy 1994; *Next!* 1999).

---

6   These are only a representative sampling of Chaum's work on e-currency. He maintains an extensive list of his publications at: http://chaum.com/articles/list_of_articles.htm

The value of each unit of eCash was tied to the national currency of the jurisdiction of the issuer's bank. To issue eCash, one needed to acquire a license to operate the eCash software from DigiCash and to establish a bank account, in which to hold users' money in reserve, in exchange for transferable claims on that money in the form of eCash units. To buy eCash units, pegged 1:1 to the currency sent, users transferred national currency—USD in the USA, DEM in Germany, CHF in Switzerland, etc.—to the issuer's bank account, and the issuer created an equivalent number of eCash units and transferred those to the user's software wallet. Each unit of eCash was a redeemable, transferable, divisible, and fungible claim on a fraction of the national currency held in reserve in the issuer's bank account. The issuer kept any interest earned on the reserves.[7]

A user of eCash thus was able to email cryptographically secure strings of characters that represented some amount of national currency—either as a gift or in payment of some good or service—to a recipient who in turn could use those strings of characters to transfer the claim to subsequent recipients, *ad infinitum*.

1.B. The Byzantine Generals Problem

A critical shortcoming of eCash was that it was highly centralized—technically and administratively—and this single point of failure was subject to hacking, subversion by criminals, legal action, physical seizure, sabotage, managerial caprice, and communication outages. Even with servers in multiple locations to provide failover[8], should the primary server go down, the problem was mitigated but not eliminated, as several servers, rather than only one, still could be taken down, in order to shut down the system.

The ideal system would be so robust that the difficulty of taking it down would be so great that the party undermining

---

7 For a detailed introduction to eCash, see Levy (1994).
8 i.e., seamlessly switching to a standby computer, when a primary computer fails for any reason

it would incur severe or even prohibitive costs. Rather than rely on a single, trusted point of failure (centralized), or even several trusted parties (decentralized), the system should be distributed among thousands or millions of redundant network nodes in multiple jurisdictions. However, for such a system to be robust, there must not be a unique master transaction record under the control of any specific person or persons. Otherwise, the system would not be distributed, but merely consist of a large collection of backups. If the master record were corrupted, then the backups would be no more than copies of a corrupt record.

Figure 6
Centralized, Decentralized & Distributed Networks

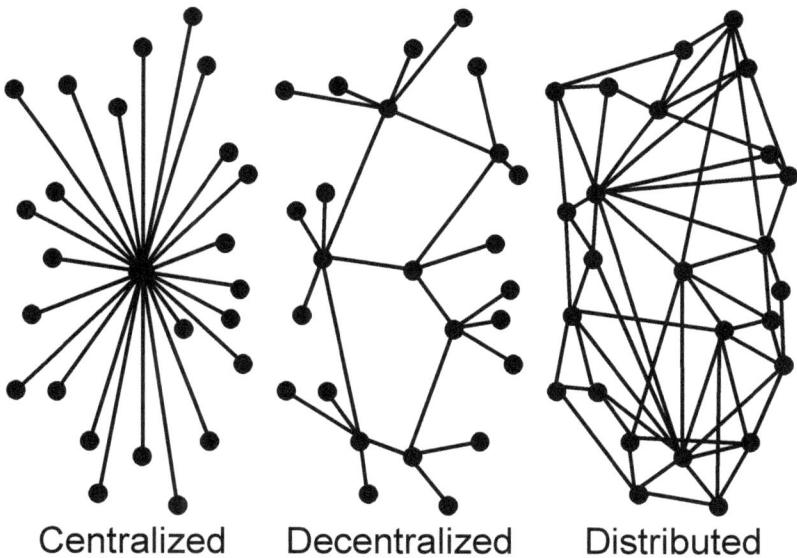

Centralized   Decentralized   Distributed

This problem became known as the Byzantine Generals Problem, first described by Akkoyunlu, Ekanadham & Huber (1975) in the context of communication between two groups of gangsters, and expanded by Pease, Shostak & Lamport (1980), and Lamport, Shostak & Pease (1982).

Very briefly, the problem concerns a group of generals in the service of the Emperor of Byzantium, each of whom commands a division of the Imperial Army. They have laid siege to an enemy city, and each general communicates with the others by messenger, which is time-consuming, discoordinated, and subject to subversion. Each general first must observe the enemy and then agree on and carry out a common plan of action with the other generals. One problem is that some of the generals might be in collusion with the enemy and are trying to subvert the loyal generals' ability to reach an executable agreement; or, some might be opportunists, who want the other generals to expend resources, while the traitors hold back, so that the traitors can avoid expending men and materiel, and perhaps even seize the loyal generals' lands at the end of the war.

Each general is aware of the situation, and each has an incentive—whether or not he is in collusion with the enemy—to harbor his men and materiel. However, all also know that, if they do not act in concert, then all will lose at the hand of the enemy.

The generals need a protocol to ensure that all loyal generals will decide upon the same plan of action and carry it out. All loyal generals will carry out faithfully the plan that they agree on, but traitors' actions will be unpredictable. The protocol must guarantee that loyal generals will carry out their duty regardless of what the traitors do.

The Byzantine Generals Problem addresses the question of the minimum proportion of traitorous generals, who can cause loyal generals to adopt a bad plan. In the realm of computer science, the generals are nodes in a distributed computer network, the Emperor is the purpose of that network, the messengers are communications sent among nodes, and traitors are nodes that either have been compromised or simply are malfunctioning.

For the quarter-century before Nakamoto (2008) released the Bitcoin white paper, it was widely accepted among mathematicians and computer scientists that the minimum number of the total number of nodes (n) that are traitorous (t) needed to disrupt the distributed network irreparably is surprisingly small: $t = (n-1)/3$. If n is very large, then one would need to subvert approximately 1/3rd of the nodes—and not a majority or some super-majority—in order to take down a distributed network; and, in a four-node network, one would need to subvert only one node (1/4th of the total). (Akkoyunlu, Ekanadham & Huber 1975; Pease, Shostak & Lamport 1980; Lamport, Shostak & Pease 1982)

This meant that any distributed computer network would require trusted operators for every node. This reduced the number of potential operators, since each would have to be vetted and monitored. This also meant that a distributed variant of eCash was inexorably subject to compromise.

1.C. Blockchains

Nakamoto (2008) demonstrated that the Byzantine Generals Problem is not insoluble, and that a secure, distributed, and above all *trustless* system is possible. This involves a public transcript of all messages that have been transmitted within the network since its inception, plus a procedure that enforces a rule allowing only one node at a time to transmit a message to the network.[9]

Every node in a cryptocurrency system has a copy of the entire transaction history, which is updated approximately every ten minutes. If any node's copy does not match the other nodes' copies, then that copy is ignored and replaced with a copy that matches the majority's copies. The version of the blockchain that is embraced by more than 50% of the nodes is deemed to be the true blockchain. The only way to subvert the

---

9   The description that follows is a simplification for a non-technical audience that corresponds with the very early days of the Bitcoin network. In the meantime, the maintainers of Bitcion's reference implementation have adapted users' access to the network to ensure communication efficiency. Specifically, not all user interfaces have full copies of the transaction ledger (blockchain)—although, anyone who wants to can access a copy—and only a small subset of users (miners) now process blocks of transactions.

system would be to have more than half of the nodes corrupt their transaction histories in precisely the same way. Conventionally, this is referred to as a 51% Attack, although the number is actually 50%+1.

However, if one did assemble enough computing power to corrupt the system in this way, then some if not all users would lose faith in it. Rational users would abandon it, which potentially would render the value of the equipment of the owners of the corrupted nodes worthless. Thus, economic incentives are expected to prevent a costly 51% Attack.

Hard-coded into the Bitcoin software is a cap on the total number of units in circulation at approximately 21 million[10], each divisible to the 1/100 millionth (satoshi). Rather than serve as claims on some other asset, as was the case with eCash, bitcoins can be seen as units of a new category of commodity in their own right. Their scarcity is ensured by military-grade encryption and by free and open source software that is continually audited by developers. Bitcoins are divisible to such fine granularity that, if the global population were 10 billion, every individual in the world could hold 210,000 satoshis on average. They are presumed to be fungible, although the fact that the transaction history is freely available means that one might distinguish among bitcoins by their previous ownership. However, currency notes have unique serial numbers and bank account records are similarly traceable, and we consider, e.g., US dollars to be fungible. And, because they exist in electronic form, they are durable and portable, so long as at least one node is running somewhere in the world.

1.C.i. Proof of Work

Bitcoin and similar cryptocurrency systems replace eCash's centralized servers with relatively small pieces of software that run nodes on each user's computer. The software restricts the ability of each node to communicate through a process known as Proof of Work (Dwork & Naor 1993; Gabber, Jakobsson,

10   The actual number is 20,999,999.9769 bitcoins.

Matias & Mayer 1998; Jakobsson & Juels 1999) that involves being the first to solve a fiendishly difficult computational problem, before being able to transmit confirmation of the integrity of the most recent batch of transactions (block) to the network. Because nodes have copies of the transaction history of the entire series of blocks (blockchain) from the system's inception to the most recent block, it is as though everyone were able to look over everyone else's shoulder to make sure that no one is cheating.

In a Proof of Work (PoW) system, each node confirms each transaction as it is broadcast through the network, and the transactions are collected into *blocks* of transactions. Once a block has been completed and broadcast, each copy of the node software calculates the hash of that block with a string of random data (nonce) appended to it, looking for a hash that begins with some number of zeros. Each node appends its own random nonce to the block, calculates the hash, and checks for a zero at the head. If the first character is not a zero, the software generates a different random nonce, and tries again... and so on, until one of the nodes on the network gets lucky, finds a hash that begins with a zero, and announces its success. The nodes confirm that the specific block-plus-nonce announced indeed yields a hash that begins with a zero, and then begin working on the next block. This process is referred to as *mining*.

Anyone can operate cryptocurrency mining software. It is not feasible to exclude any miner from the network. Miners operate 24/7 worldwide, with concentrations in areas where energy, equipment, and Internet bandwidth costs are relatively low. Because of the heat generated by computer processors, favored locations tend to be as far from the Tropics as possible.

The difficulty of the calculation is set by how many zeros must appear at the beginning of the hash. In early 2009, the first node to calculate a hash with a single zero at the head received a reward of 50 bitcoins. Once the mining reward has been awarded to the winning node, the race begins again on the next block of transactions that were completed while the nodes were searching for the previous hash with a zero at the head.

As more nodes join the system, the likelihood that someone will find a hash with a zero at the head in the next moment of time increases, and the delay between blocks shrinks. The system is designed to keep the average time between blocks as close to ten minutes as possible (144 times per 24 hours). When that time falls significantly below ten minutes—because of the increased computing power that has joined the network—the system increases the difficulty by requiring that two, three, or more zeros be at the head of the winning block. If nodes leave the network, reducing the computing power dedicated to mining, and the time between blocks rises significantly above ten minutes per block, the system reduces the number of zeros required at the beginning of the winning hash. (Antonopoulos 2015; Bitcoin Wiki 2018a)

1.C.ii. Mining Rewards and Transaction Fees

The number of bitcoins awarded for each successful block is halved every 210,000 blocks, roughly every four years. The first halving of the mining reward occurred on 28 November 2012 from 50 to 25 bitcoins per block. The second occurred on 9 July 2016 from 25 to 12.5. The third is expected in mid-June 2020 from 12.5 to 6.25. Sometime in 2140, the mining reward is expected to fall to zero, after the last of the approximately 21 million bitcoins will be released into circulation.

In addition to the mining reward, miners can charge transaction fees that are denominated in bitcoins. The transaction fee that a miner charges is at the miner's discretion. If the miner charges a fee that is significantly lower than what other miners charge, then it leaves money on the table; if it charges a fee that is significantly higher than what other miners charge, then it prices itself out of the market. The profit-maximizing fee is the equilibrium market consensus fee.

1.C.iii. Proof of Stake

The forensic economist who has been hired as an expert on cryptocurrency should avoid being pulled down a rat hole by taking the bait during cross-examination to testify on technical issues. Nonetheless, it is critical that he have a high level understanding of some of the relevant issues that could come up, so that he can anticipate when to defer to technical experts. In that spirit—and in the same way that he should at least understand what public key cryptography is—he should have a high-level understanding of Proof of Stake (PoS) and how it differs from PoW.

PoS is a family of consensus algorithms that attempt to institute formal—and thus predictable—governance *ex ante*, rather than trust that the optimal governance structure will emerge spontaneously, as is the case with a PoW system. Under PoW, anyone can bring as much computing power to the network as one wants, and it is impossible to regulate who controls what percentage of the total computing power devoted to mining on that blockchain. Even more important, miners generally do not hold cryptocurrency in proportion to their share of the mining power. Much more commonly, those who hold large cryptocurrency balances are not miners; instead they are at the mercy of the miners to behave honorably. One way of looking at it is to see cryptocurrency holders in a PoW system as owners of *capital shares* in the enterprise, and the miners as owners of *voting shares*.

PoS is an approach based on the assumption that those who have the most 'skin in the game'—not only miners with their physical capital investments but the economic majority of token holders with their financial capital investments—have the strongest incentive to ensure the integrity of the system. One way of looking at it is to see cryptocurrency holders in a PoS system as owners of shares in a newly incorporated shell company that represent both a proportional ownership stake in the enterprise's capital with voting power in proportion to one's ownership stake. PoS systems include BitShares, EOS, and Steemit.

In a PoS system, users take turns performing the role of 'miner'. Depending on the bylaws of the system, all holders might be eligible to be elected to serve in a group that is roughly analogous to a board of directors, in which 'directors' take turns confirming the validity of each block of transactions and receiving the 'mining' reward for that block. Alternatively, the board might consist of the users with the largest holdings, who either receive the 'mining' reward or volunteer their services. It is also possible to rotate the 'miner' role, randomly or deterministically, through all users, analogously to being called for jury duty.

PoS is based on the assumption that whoever performs the role of 'miner' should be known and endogenous to the system, rather than potentially anonymous and exogenous to the system. This distinction notwithstanding, PoS follows the same basic outline as Pow (Nakamoto 2008, §5 p. 3-4):

1) New transactions are broadcast to all nodes.
2) Each node collects new transactions into a block.
3) Each node works on finding a difficult proof-of-work for its block.
4) When a node finds a proof-of-work, it broadcasts the block to all nodes.
5) Nodes accept the block only if all transactions in it are valid and not already spent.
6) Nodes express their acceptance of the block by working on creating the next block in the chain, using the hash of the accepted block as the previous hash.

The notable exception with PoS is that, instead of engaging in the probabilistic PoW race run by competing mining pools in Steps 3) and 4), nodes are selected from among users as described above.

2. Bitcoin

Since its inception in early 2009, Bitcoin has grown from a small-scale computer science experiment into a global financial infrastructure that is attracting hundreds of millions of dollars of investment worldwide each year in services based on it. However, even as it gains legitimacy among regulators, central bank officials, and investors, questions persist about what Bitcoin fundamentally is: commodity, base money, currency, database, transaction history, etc.? One can view it as any or all of these things, depending on context. Descriptions based on conventional categories should be seen as metaphors, rather than definitions.

However one describes it, Bitcoin specifically and cryptocurrency generally is useful for small-scale, cross-border transactions, developing software that incorporates the automatic spending and receiving of value—*smart contracts* (Szabo 1997)—and circumventing currency controls in authoritarian regimes like Cuba or Venezuela. This is particularly valuable for the estimated 85% of humanity worldwide—including 20-25% of US residents—who are either unbanked or underbanked.[11]

This section discusses the fundamental nature of money and how it relates to cryptocurrency, monetary separation, ephemerality versus intangibility, Bitcoin in practice, advantages of Bitcoin, and drawbacks of Bitcoin.

---

11 Granted, cryptocurrencies can be used for illegal activity, as can cars, guns—which are protected by the Second Amendment—knives, telephones, cash, and many other common items.

2.A. Floor Wax *and* Dessert Topping[12]

Perhaps the most basic question about Bitcoin that a forensic economist is expected to answer is, "What *is* Bitcoin?" As indicated above, and discussed below, it is a software application that implements a protocol for the distributed maintenance of a blockchain. Because that is likely to lead to exponentially more questions, rather than settle the matter, we tend to fall back onto metaphors, in the hopes that this will clarify the situation. Which metaphor one chooses can have a profound impact on the implications of one's testimony.

Aside from the fact that metaphors are, by their nature, imperfect explanations or definitions, forensic economists face the potentially more nettlesome fact that different metaphors imply different legal statuses and regulatory regimes. Sometimes—from the perspective of accounting, economics, and finance—the distinction among different metaphors is as meaningful as it would be if the statements **4+5=9** and **3x3=9** were governed by different sets of statutes, regulations, and administrative rules that were overseen by the Ministry of Addition on the one hand and the Ministry of Multiplication on the other, and that regulators in each Ministry were jealous of their turf, and maintained the distinction zealously, whereas the accountant, economist, or financial analyst sees them as equivalent and interchangeable, because they both equal **9**.

This is relevant, when regulators, prosecutors, and judges dwell on the distinction between *money* on the one hand and *commodity* or *property* on the other hand, in order to determine whether the dispute involves a financial crime. This point is often the crux of the dispute being decided. It is crucial that the forensic economist testifying as an expert on cryptocurrency be very clear in his mind, what his position here is, in part because financial crimes often carry more severe penalties

---

12 The title refers to a *Saturday Night Live* skit from the 1970s about a fictitious product called Shimmer that was presented as both a floor wax and a dessert topping. One should be able to find video or audio copies of the skit on YouTube.

than property rights violations or even violent crimes. While one might appreciate the ambiguity of the false dichotomy between money *xor*[13] commodity or property from the perspective of economic theory—especially, the implications of *monetary separation* discussed below—judges, juries, and lawyers tend to want clear, monosyllabic answers that enable them to decide which specific statutes, regulations, or administrative rules are relevant to the case.

The convention in accounting and finance is that *money* and *numeraire* are equivalent in practice, although they are different in economic theory. For example, if one US dollar is defined as $1/35^{th}$ of an ounce of gold, then the money (gold) is a tangible thing and the numeraire (dollar) is an intangible number. Nonetheless, in daily usage, this distinction is ignored, and a dollar is treated as a dollar.

If prices are quoted in dollars, dirham, euros, lira, pesos, rubles, rupees, yen, yuan, or whatever else, then *that* is the money and the bitcoins are the goods being priced in that money. If, conversely, one exchanges bitcoins (BTC) directly for, e.g., ethers (ETH)[14] or some other good or service, bypassing 'money' altogether, then that exchange is *barter*, which might be regulated by a separate set of statutes, regulations, and administrative rules from those that govern *sale* and *purchase*. State, federal, and US Army courts have accepted the *commodity* metaphor, following Guidances issued by IRS and CFTC officials, albeit a commodity or form of property that its creators *intended* to be used *as money* (in barter).

The forensic economist who is serving as an expert witness on cryptocurrency should make every effort to stand firmly within the sandbox of his specific expertise, and not offer opinions—*especially* legal opinions—that fall outside his sandbox. That said, as an expert witness, alerting the court to CFTC, Federal Reserve, FinCEN, IRS, SEC, Bank for International Settlements (BIS), IMF, World Bank, or other Central

---

13 The term *xor* refers to the *exclusive or* from Logic; e.g., **A xor B** means *either* **A or B**, but *not* **A and B**. You get one, but not both.
14 another popular cryptocurrency

Bank or Finance Ministry opinions that have influenced one's analysis and conclusions *is* within one's purview.

2.A.i. What Is Money?

Lawyers and judges, who do not deal regularly with fundamental economic questions are often incredulous, when the topic of money inevitably comes up in a cryptocurrency case. Money is one of those things that seemingly everyone has an opinion about, and yet the vast majority are at a loss, when asked to define *money* with precision and specificity. The natural next step is to consult the relevant statutes, which generally are silent on the definition of *money*, in contradistinction to *currency*, *qua* a *claim* on money that is "issued by a government." The step after that is to consult *Black's Law Dictionary*, which yields something like:

> **Money**: A general, indefinite term for the measure and representative of value; currency; the circulating medium; cash. "Money" is a generic term, and embraces every description of coin or bank-notes recognized by common consent as a representative of value in effecting exchanges of property or payment of debts... Money is used in a specific and also in a general and more comprehensive sense. In its specific sense, it means what is **coined or stamped by public authority**, and has **its determinate value fixed by governments**. In its more comprehensive and general sense, it means wealth.
> (Black, H.C. 1910) [emphasis added]

Although this might seem to settle the question, it is not unusual for the term crypto*currency* to inspire a line of questioning that leads one deep into the prehistory and fundamental metaphysical nature of money as a social institution. Depending on the forensic economist's background and intellectual interests, answering truthfully and completely might imply the invocation of terms like *ontological* and *epistemological*, which, though accurate are unhelpful. How one gets around this is a matter of personal style. The primary point here is that the forensic economist serving as an expert witness on cryptocurrency should be prepared for this line of questioning.

2.A.ii. Monetary Separation

Historically, money was understood to be anything that served four fundamental functions:

- **Medium of Exchange**: The thing that is universally accepted in exchange for all other goods and services.
- **Store of Value**: Something that is expected to buy at least as much or many goods and services in the future as in the present.
- **Unit of Account**: The accounting and bookkeeping numbers that we use to record transactions and debts.
- **Measure of Value**: The economic standard for what things are worth.[15]

The ideal money has the following qualities:

- **Durable**: It does not evaporate, crumble, or rot.
- **Fungible**: Every unit is economically identical to every other unit.
- **Divisible**: It can be subdivided into smaller portions, unlike cows, paintings, tools, etc.
- **Scarce**: The supply is constrained, unlike sand or water.
- **Portable**: This is implied by scarcity, but excludes, e.g., huge boulders, that are scarce but hard to transport.
- **Recognizable**: Users can identify it easily, accurately, and consistently.

---

15 For example, we might say that a cow is *worth* $2,000 and that a car is *worth* $40,000, but we would not say that a car is *worth* 20 cows.

Monetary separation is the performance of different functions of money with different media. For example, under the gold standard, bars of gold in a vault served as the store of value; paper currency notes and base-metal coins served as the medium of exchange; the dollar was the unit of account and the measure of value, whether the dollar was defined as $1/20^{th}$ oz. or $1/35^{th}$ oz. of gold. With Bitcoin, the bitcoins are the store of value and the medium of exchange, but not the unit of account or the measure of value.

In this way, one might argue that a bitcoin is as much money as a dollar is, because it performs only some of the functions of money, as the dollar does. Alternatively, one might argue that a bitcoin is the opposite of a dollar, because a dollar serves as the unit of account and measure of value, but not the store of value or the medium of exchange, while a bitcoin serves as the store of value and medium of exchange, but not the unit of account or the measure of value.

These questions strike at the heart of monetary economics and jurisprudence related to money, banking, and finance. The answers are not obvious in either realm of inquiry and are subject to economic and legal interpretation. The forensic economist is the economic expert explaining all this to legal experts, who then apply it in civil and criminal trials. He must be confident in his analysis and ensure that it has a solid foundation. Above all, he must be prepared to defend his position. That is the nature of expert testimony. If any of this were self-evident, expert witnesses would be redundant.

That there is wide disagreement in the field of monetary economics is unsurprising. The literature on money dates at least as far back as Aristotle (Meikle 1994; Horace White 1902; Menger 1892). An author—whose work on the subject was widely known in the days leading up to the creation of the Federal Reserve System in the USA in 1913—was Horace White (1902), who observed that *money is a commodity that individuals accept voluntarily in exchange for all other goods*. He distinguished between *real money* and *promises to pay*, noting that the difference between money and currency is analogous to the difference between a meal and a meal ticket, or a coat and a coat check; today, we might include the difference between a car and a valet ticket. The former is a thing of value, and the latter is a claim on that thing.

Fama (1980, 39) draws a related distinction: "[C]urrency and an accounting system are entirely different methods for exchanging wealth." It is one thing to hand someone a currency note, and another thing to reassign an account receivable, which is the essence of the payment network that is built on the banking system.

Monetary separation has been endemic since the creation of paper notes—which are *promises to pay* the bearer some amount of *real money*—and even more so with the invention of fractional reserve banking, in which the store of value is not a bar of gold, but instead a loan owed to the bank. It was taken to its logical extreme, when Federal Reserve Notes were declared in 1913 to be legal tender in the USA and traded at parity with gold certificates. Later, it became a federal crime in 1933 for US persons to own the US dollar (USD), defined as $1/35^{th}$ of one troy ounce of gold. Finally, in 1971-1973 last vestiges of the USD to gold were severed, and it has become a *promise to pay* $1.00, backed by a loan with a face value greater than $1.00 owed to the Federal Reserve. In this final incarnation, one might conclude that monetary separation is *total*, now that the USD is a *promise to pay* USD, itself being a *promise to pay*. One might recognize this triumph of pure fiat over gold as money as the establishment of a new category of self-backed financial asset. In this way, the *medium of exchange*, the *unit of account*, the *measure of value*, and the *store of value* are all the USD.

2.B. Bitcoin in Practice

Cryptocurrency systems, like Bitcoin, are based on Public Key Cryptography (described in §1.A.i.a. "Public Key Cryptography")—which involves the use of public/private cryptographic key pairs—and Hashing Functions (described in §1.A.i.c. "Hash Functions"). Cryptocurrency transfers require the holder of a private key that grants the holder access to a cryptocurrency address that is deterministically associated with the corresponding public key of that key pair to initiate a transfer with that private key. A private key—e.g., x9y8z7—serves the same purpose as a password. Let us say that x9y8z7 is the private key (think: 'password') that is associated with Bitcoin address (think: 'account') abc123. Anyone can spend bitcoins to abc123, in the same way that anyone can transfer money into one's bank account, but only the party that knows the secret "x9y8z7" can use a Bitcoin wallet to instruct the system to transfer bitcoins out of abc123 and to a different address.

Bitcoin addresses are strings of random characters, e.g., 1HVGxZRkbQNyKqdg4ECFwZqcvkzKCX9bZZ. (This is an actual Bitcoin address that corresponds with the illustrative "abc123" in the previous paragraph. Only the owner of that account knows what the corresponding private key is.) It is not possible to recognize directly which individual or organization controls a specific Bitcoin address, but it is possible to trace a chain of transfers that either has originated at or ends at a known party. Utilities that enable the tracing of Bitcoin transfers are available online, including Blockchain.info[16][17], which was launched in August 2011. Services exist that enable the tracing of Bitcoin addresses to known persons, including Chainalysis[18], an anti-money laundering service provider that works with financial institutions, merchants, and regulatory and law enforcement agencies worldwide.

---

16 http://blockchain.info
17 The reader can search on Blockchain.info for the Bitcoin address:
    1HVGxZRkbQNyKqdg4ECFwZqcvkzKCX9bZZ
18 https://www.chainalysis.com/

One can treat Bitcoins as irredeemable units of a new category of intangible commodity in their own right. Their scarcity is ensured by military-grade encryption and by free and open source software that is available to all and continually audited by developers. Bitcoins are presumed to be fungible, although the fact that the transaction history is freely available means that one theoretically could distinguish among bitcoins by their previous ownership. However, this is not the practice among users, regulators, or law enforcement officials. Instead, each bitcoin—or, rather, each satoshi—is treated as fungible, even though their movements can be traced through the public transaction ledger (Meiklejohn, Pomarole, Jordan, Levchenko, McCoy, Voelker & Savage 2013; United States District Court for the Northern District of California 2015, Exhibit B). This is similar to how USD notes are treated as fungible, even though each bears a unique serial number and could be tracked as it entered and left the banking system, but generally is not.

Hard-coded into the Bitcoin software is a cap on the total number of bitcoin units in circulation of 21 million, each divisible to the $1/100$ millionth (satoshi). Thus, bitcoins are divisible to such fine granularity that, if the global population were 10 billion, every individual in the world could hold 210,000 satoshis on average.

As discussed briefly in §1.C.i., the first batch of fifty bitcoins was released on 3 January 2009 at a rate of fifty new bitcoins approximately every ten minutes thereafter. The software reduces the rate of release by half, approximately every four years. The first time, to twenty-five bitcoins every ten minutes, was in November 2012. The next halving is expected sometime in 2016 to 12.5. This halving process is expected to continue until approximately 2040, after which the cap shall have been reached, and no more new bitcoins are to be released into circulation. The only way to change this limit is to get users who collectively control more than 50% of the computing power in the Bitcoin system to agree to run modified software with a different cap. However, doing so could undermine market confidence in the value of the system as a whole, as well as the value of each bitcoin unit, which would render the specialized computers that confirm Bitcoin transactions worthless. (Bitcoin Wiki 2018b)

Experience over more than a decade has revealed that the Bitcoin system and the blockchain that it maintains can be used for much more than just transferring bitcoin units among users. Bitcoins can be seen simultaneously as virtual currency deposits, non-voting capital shares, and a new kind of commodity that can be used as a medium of exchange. Developers have found ways to embed messages into the Bitcoin blockchain, which time-stamps them such that they can be seen as a kind of 'notarization' (Kirk 2013). Others have created ways to tag specific satoshis with extra strings of data (colored coins) that give each series a unique 'color'—like a brand on cattle—that can be used as proxies for or claims on shares of equity, bonds, future or forward contracts, options, leases, annuities, or any other financial assets (Rosenfeld 2012). One can use colored coins as warehouse receipts for assets held in reserve, thereby creating 'virtual gold', 'virtual dollars', etc. One can use one's Bitcoin software (wallet) to create digital signatures of blocks of text, including contracts. It is also possible create Bitcoin accounts that require more than one private key to open them, as a form of escrow that can be used in lieu of banker's acceptances and letters of credit. And, developers continue to find cleverer and more unexpected uses for this platform that can be used from anywhere in the world with Internet access, for all manner of transactions and services that are *de jure* regulated and now have become *de facto* unregulatable.

2.B.i. Ephemerality

In addition to whether or not bitcoins are money, if one determines that bitcoins are commodity units or property, the question of whether they are *tangible* might come up. For example, if the defendant is charged with *larceny*, then the issue of tangibility versus intangibility becomes the crux of the dispute, because larceny generally involves tangible property.

As discussed above, historically money performed the functions of medium of exchange, store of value, unit of account, and measure of value. Cronin (2012) points out that nothing serves all four functions anymore, and that monetary functions have been separated from each other. Additionally, the relationship between *money* and *currency*—prior to institution of pure fiat central bank money in the 20th Century—was understood to be similar to the relationship between a car and a valet ticket (White 1902).

Although Graeber (2014) challenges the historical accuracy of Menger's (1892) story of the origins of money, as a conceptual model it serves our purposes here. At the dawn of civilization, when humans began living in communities that were too large for everyone to be related directly to—or even acquainted with—everyone else, the institution of money in the form of gold and silver coin emerged. Later, money holders stored their money with treasuries, and later banks, and used their warehouse tokens as currency in transactions. Beginning in the Late Renaissance, bank owners began to organize central banking networks, and fractional reserve banking led to the circulation of currency backed by promissory notes rather than by money. By the 20th Century, pure fiat began to circulate at parity with gold- and silver-backed Treasury Notes. Finally, in the early 1970s, all ties between currency and underlying gold and silver money were severed (Bernstein 2008), and all national currencies began to float against each other in the global foreign exchange market.

At each stage along the way, the circulating medium has become increasingly intangible, to the point where each USD, EUR, JPY, etc. is backed by a promise to repay loans denominated in those same currencies.

Bitcoin represents a step beyond mere intangibility into *ephemerality*. Conventionally, intangible assets represent some underlying asset; options and futures are contracts to buy or sell underlying assets, patents represent inventions, copyrights are associated with texts, and goodwill is the excess value of a going concern. In each case, the intangible asset can be seen as a derivative, in the broadest sense.

Continuing this developmental process, bitcoins exist *in se*, representing no claim on any underlying asset. In this way, they are like a new category of commodity that has no physical existence. Like money, bitcoins are treated as fungible, even though the freely available transaction history enables anyone to distinguish among bitcoins by their previous ownership (Meiklejohn, Pomarole, Jordan, Levchenko, McCoy, Voelker & Savage 2013; United States District Court for the Northern District of California 2015, Exhibit B).

2.B.ii. Nonexistent Buckets of Sand

One way to look at Bitcoin is to imagine that one had 21 million eight-gallon (30 liter) buckets in a secure warehouse; that each bucket contained 100 million grains of sand; and that each grain had been etched with a unique serial number. Now, imagine that some individuals, for whatever reason, decided to use these uniquely numbered grains of sand as placeholder tokens, like Monopoly™ money or poker chips. Imagine, further, that the sand were locked away, and that transfers of ownership were effected by bookkeeping entries, rather than physical delivery. Every transfer of control of each grain of sand from one user to another would be recorded in this transaction history. Anyone who wanted to have a copy of it could have one at any time. If anyone's transaction history differed from the record recognized by the majority of users, it would be discarded and replaced with a copy of the record recognized by the majority.

Approximately every ten minutes, all of the most recent transaction records would be collected and validated by the users who maintained copies of the transaction history[19]. Then, each user would try to solve a fiendishly difficult mathematical

---

19   In the spirit of economic models, this is a simplification that reflects the early days of Bitcoin, but no longer applies as described. Today, users and miners use different software that is optimized to each group's needs.

puzzle that were easily confirmed as correct once solved. The purpose of this step would be to maintain the rate of approximately one block of transactions every ten minutes, because it should take that long on average for someone to solve the puzzle. If the success rate were faster, then the difficulty of the puzzles would increase, and *vice versa*. The first to complete this last step would get the reward for that 'block' of transaction validations, which would be appended to the bottom of the transaction history, known as the 'blockchain', and the cycle would restart. This process of transaction validation and puzzle-solving would be referred to as 'mining'. The purpose of the mining reward would be to create an incentive for users collectively to ensure the integrity of the transaction history, rather than trust a corruptible central authority. If the recipient of this mining reward of some buckets of sand did not hoard the sand, but instead gave it away, used it to buy goods and services from merchants who accept it, or sold it, the newly released buckets of sand would come into circulation.

Now, ignore the buckets and the sand. They do not exist. Keep the transaction history, though. That is Bitcoin.

2.B.iii. There's No Accounting for Bitcoin[20]

If—instead of consisting in a globally distributed network of anonymous nodes with each running a copy of the same software application—Bitcoin were a financial services firm, with a CEO, headquarters, etc., that provided precisely the same services, one can see a unit of bitcoin as a non-voting share of the system's capital stock. In fact, some refer to Bitcoin specifically and cryptocurrencies generally as Distributed Autonomous Companies (Economist 2014; Larimer 2013).

Bitcoin's primary service is keeping track of users' balances of bitcoins and confirming transfers of bitcoins among users. As discussed above, it also provides many other services, including the time-stamping of documents as a kind of 'nota-

---

20 This section is based on Pluzhnyk & Evans (2014), presented at the 2016 annual meeting of the Financial Education Association.

rization' (Kirk 2013), creating tokens that can represent financial contracts (Rosenfeld 2012), and the replacement of bankers' acceptances and letters of credit with multi-signature wallets (Evans 2015). If the services that Bitcoin provides were provided by a conventional financial services firm, what follows would be unexceptional. The salient difference is that these services are provided by a network of millions of users worldwide, who run Bitcoin software on their computers that automates all of its functions, with no CEO, no headquarter, and none of the other legal trappings of a conventional firm.

When one considers that Bitcoin's market capitalization at the time of this writing is larger than more than 50 of the firms in the S&P 500 (Chicago Board Options Exchange 2019; CoinMarketCap 2019)—and that it operates as a Distributed Autonomous Company (Economist 2014; Larimer 2013) that has no headquarters, no executives, and no employees—it provides an example of a more extreme conundrum that Accounting and Finance instructors face, when applying conventional analytical tools that were developed for a manufacturing-based capital/labor economy to information-based firms in a knowledge/service economy. Today, new great fortunes are made by the founders of firms with relatively small physical capital bases that do not resemble the Acme Widgets that features prominently in textbooks. With pure service firms, accounts receivable, inventory, and short-term and long-term debt are effectively zero. In such a world, conventional ratio analysis becomes difficult, and often meaningless. When we go a step further into a world in which services are driven by automated smart contracts (Miller & Stiegler 2003; Szabo 1997), the notional balance sheets of such 'firms' are largely zeroed out, even though their equity values are potentially very large.

Market participants' expectations of the value of future services enabled by Bitcoin (F), the time until those services become available (T), and the consensus risk-adjusted discount rate (r) determine the market value of a bitcoin (P), such that:

$$P = \frac{F}{(1+r)^T}$$

Sometimes these expectations change dramatically over short periods due to uncertainty associated with innovation adoption (Moore 1999; Rogers 1962) and noise trading (Black 1986). Over long periods, the trend is generally upward, as market participants' expectations of Bitcoin's advantages (F) increase, the expected time until those advantages are realized (T) shortens, and required returns (r) decrease as uncertainty and perceived risk are mitigated by experience.

Notwithstanding CFTC (2014) and IRS (2014) determinations that bitcoins should be treated as commodities, one can see them as non-voting capital shares (Economist 2014; Larimer 2013). Indeed, classifying them as commodities presents potential problems. Foremost, bitcoins have no physical existence; their only manifestation is in the public ledger that keeps track of them (Evans 2015). They also do not fit conventional definitions of intangible assets, which are associated with some underlying asset, excess value in a firm (goodwill), an underlying financial asset (derivatives), inventions (patents), texts, videos, or images (copyright), exclusive geographic territories (franchises), etc. Analogies to ephemeral commodities like electricity similarly fail in the details, as electricity is consumed, and bitcoins are not. Likewise, if one wants to accumulate electricity, one must invest in physical batteries, whereas if one wants to accumulate bitcoins, one simply hoards them.

As non-voting capital shares of equity in a distributed autonomous company, bitcoins present an unusual, though not inconceivable, financial asset. They can be seen as fractional claims on the value created by the Bitcoin system in excess of creditor claims. That value is observable in the market for bitcoins that has operated 24 hours per day, every day since the first public exchanges (discussed in detail in §4.) began operating in 2010 (Bitcoin Wiki 2015), with no breaks for weekends or holidays. These 'shares' are non-voting, as Bitcoin users' opinions are not sought by the other key stakeholders within the system: miners, node operators, developers, investors and founders of projects based on Bitcoin, etc. However, their value is in proportion to the fraction of the total value of the Bitcoin system that each bitcoin holder controls.

In conventional terms, Bitcoin's balance sheet looks like this, where XBT here is the most recent bitcoin price multiplied by the number of bitcoins in circulation (market cap):

Figure 7
The Balance Sheet of the Bitcoin System

ASSETS		LIABILITIES	
Cash	0	Accounts Payable	0
Accounts Receivable	0	Notes Payable	0
Inventory	0		
Fixed Assets	0	Long-Term Debt	0
		Paid-In Capital	0
GOODWILL	**XBT**	Retained Earnings	**XBT**
TOTAL ASSETS	**XBT**	**TOTAL LIABILITIES**	**XBT**

Before Bitcoin was first posited, functioning software was released, a market in bitcoins emerged, and investors invested more than $1 billion in Bitcoin ventures (CoinDesk 2016), this might have been an intellectual curiosity. However, now that Bitcoin's market capitalization would qualify it for entry into the S&P 500 if it were a conventional firm, we are faced with the very real question of how accountants and financial analysts should model distributed autonomous companies like Bitcoin.

2.B.iv. Forks

A cryptocurrency fork is functionally similar to a corporate spinoff, in which the executives of a firm create a "daughter firm" from an existing division or combination of assets of the "parent firm" and distribute shares in the daughter firm to the parent firm's shareholders in proportion to their holdings. In this way, if a shareholder owned 10% of the parent firm before the spinoff, that shareholder now would own 10% of the shares of the parent firm and 10% of the daughter firm that had been spun off. Examples include Dean Witter, which was spun off from Sears Roebuck & Co. (Sears) in 1993, and Allstate, which was spun off from Sears in 1995. Prior to the spinoffs, Sears shareholders owned the assets that comprised Sears and what eventually would be two new daughter firms; after the spinoffs, the shareholders owned shares in three different firms.

When a cryptocurrency is forked, holders gain access to units of the new cryptocurrency in proportion to their holdings in the original cryptocurrency as of the date and time of the fork. Essentially, a new software system's transaction history begins with a copy of the transaction history of the cryptocurrency that it is forking from, giving it a ready-made user base, rather than start *ab ovo*.

At the time that this text was written, more than two dozen forks of Bitcoin with market valuations had been issued. (See "List of Bitcoin Forks" at ICO Now[21] for a list of Bitcoin forks and CoinMarketCap[22] for the prices of the ones that are listed on cryptocurrency exchanges[23].)

During a fork, someone clones a copy of the Bitcoin software with the existing keys and addresses intact. The private keys do not go anywhere. They are just strings of characters. If one is using the existing Bitcoin software, then one can use x9y8z7 from the illustration above to initiate a payment out of abc123 in the Bitcoin network. If one runs a copy of the wal-

---

21 https://iconow.net/list-of-bitcoin-forks/
22 https://coinmarketcap.com/
23 Cryptocurrency exchanges are discussed in detail in §4.

let software that communicates with the forked system—e.g., Bitcoin Cash or Ethereum Classic—then one uses the same x9y8z7 to access the address abc123 in that system.

2.B.iv.a. The Pottersville Fork of Bedford Falls

One way of looking at a fork is to imagine that someone had made an exact replica of Bedford Falls from *It's a Wonderful Life*[24] just before George Bailey was conceived. Every person, house, rock, tree, squirrel, etc. in Bedford Falls would be replicated down to the finest detail in the new town; the group 'forking' Bedford Falls might call their copy Pottersville. Keys from Bedford Falls would open locks in Pottersville, and vice versa. Anyone who owned a house in Bedford Falls at the time of the fork would own the corresponding house in Pottersville. Immediately following the 'fork', the two towns would be exactly identical.

Following the 'forking' of Bedford Falls into two post-fork towns, any changes in Pottersville would not be reflected in Bedford Falls, and vice versa. For example, one could sell one's house in Pottersville and used the money to improve one's house in Bedford Falls. Each town would take on a life of its own in the moments following the 'fork', and they would resemble each other less and less over time. The differences between them might be very small at first, or they might be as large as George Bailey's being conceived, born, and reared in Bedford Falls and never appearing in Pottersville.

2.B.iv.b. The Dao of Spinoffs

Until Ethereum Classic forked from Ethereum, another popular cryptocurrency, in July 2016, the conventional wisdom among those involved with cryptocurrency—dating at least back to January 2009—was that forks would be transient, and that one

---

24 If the reader is unfamiliar with *It's a Wonderful Life*, he can read a synopsis of the plot at Wikipedia:
https://en.wikipedia.org/wiki/It%27s_a_Wonderful_Life

or the other, but never both, would continue to operate, while the other withered.

The idea was that none of the Byzantine Generals discussed in §1.B. would secede. If any were led astray, the expectation was that they would return to the fold and abandon their wayward path.

This changed, when a group of Ethereum developers opposed the unwinding of transactions on the Ethereum blockchain following the collapse of The DAO[25], an early attempt at the organizing of a quasi-distributed and unregulated investment fund built and running on the Ethereum network. The DAO raised approximately $168 million worth of *ethers* (ETH), the cryptocurrency units that circulate within the Ethereum network, in May 2016. The plan was for shareholders of The DAO to vote on projects that implement smart contracts (Miller & Stiegler 2003; Szabo 1997) running on Ethereum, and to invest in them from the pool of ETH raised from the founding shareholders (Metz 2016). In the same month, Mark, Zamfir & Sirer (2016) released a paper that detailed security vulnerabilities in The DAO's code that could enable hackers to steal ETH from The DAO and called for its suspension, until those design flaws were corrected (Popper 2016).

The DAO's organizers ignored the warnings. The next month, hackers exploited the vulnerabilities that Mark, et al., had identified and stole 3.6 million ETH (approximately 1/3rd of the total raised, worth approximately $50 million) from The DAO's Ethereum accounts. (Finley 2016)

This presented the developers of the Ethereum system with a choice a) of letting the theft stand, on the rationale that The DAO was a project that ran on Ethereum, and that it was not the responsibility of Ethereum's developers to protect the investors of The DAO, or b) rolling the Ethereum blockchain back to the moment before the theft, simultaneously erasing all subsequent transactions that were not related to The DAO. The

---

25 DAO is an abbreviation that stands for Distributed Autonomous Organization, a term coined by cryptocurrency enthusiasts who preferred not to use the term Distributed Autonomous Company (Larimer 2013).

DAO's investors and leading members of the Ethereum community—with a predictably large overlap of the two groups—vote in July 2016 to fork the Ethereum blockchain and unwind all Ethereum transactions of the previous month back to the moment before the theft.

Some Ethereum developers disagreed with this course, seeing it as a subversion of Ethereum's independence as an impartial and dispassionate transaction system, and potentially setting a precedent for calling on Ethereum to provide customer support in the future for individuals who were aggrieved by other Ethereum users. Their argument was that, if Ethereum were subject to future rollbacks, then this would create uncertainty. How could one be certain that one's Ethereum transaction would not be unwound in response to some other fiasco that one had nothing to do with? It would be as if Federal Reserve officials rolled back all USD transactions back to the moment before Bernie Madoff started his Ponzi Scheme or the moment before Enron's executives started hiding liabilities off-balance-sheet.

Those Ethereum users who embraced this worldview let their computers continue to mine the original Ethereum blockchain—that was supposed to wither and fade away, as Ethereum miners followed the new fork—renaming it Ethereum Classic.

The operators of some major cryptocurrency exchanges (discussed in detail in §4.), along with the executives of Grayscale Investments—managers of the Bitcoin Investment Trust, shares of which trade in the OTC market under the ticker symbol GBTC—recognized both Ethereum Classic and Ethereum, and a new cryptocurrency was born.

Ethereum Classic demonstrated the economic viability of a forked cryptocurrency that had been spun off, rather than started *ab ovo*. Until this time, the received wisdom was that such *forked coin* would not gain traction.

The next major forked coin was Bitcoin Cash, which forked from Bitcoin in August 2017. This fork resulted from an acrimonious and irreconcilable divergence of visions among

Bitcoin developers and users concerning technical issues that are beyond the scope of this guide. If a forensic economist is called upon to testify on issues related to the forking of Bitcoin Cash from Bitcoin, he can find vast amounts of information on-line with a few intuitive keyword searches, and he can consult with Bitcoin supporters on Reddit's **/r/bitcoin** subreddit and Bitcoin Cash supporters on Reddit's **/r/btc** subreddit. Be aware, though, the two camps are highly partisan, and finding a neutral voice could be as difficult as finding a neutral voice on Donald Trump or Hillary Clinton.

2.C. Benefits of Bitcoin

As discussed below, the current value of a bitcoin (P) is determined by market participants' expectations of three factors that vary, sometimes dramatically over short periods of time: the value of future services and cost savings (F); the required return that includes premiums for various categories of perceived risk (r); and the time until F is expected to be realized (T). The present value relationship at the heart of finance holds:

$$P = \frac{F}{(1+r)^T}, \text{ where}$$

-     **F** includes the value of the ability 1) to transact directly with the billions worldwide who do not have bank accounts but own mobile phones, 2) of individuals in regions with relatively high consumption costs to buy from online retailers who accept bitcoins at Chinese or US prices, plus shipping, rather than at multiples of those prices that prevail locally, 3) to issue transferable tokens that can represent any financial asset or contract, 4) to remit any amount between any locations that have Internet access, and at a trivial cost, 5) to hold value with greater privacy than exists with legacy stores of value, especially in areas where holders of large bank account balances are in danger of being kidnapped for ransom, and 6) any financial or non-financial innovations that Bitcoin and related technologies enable that have been prohibitively expensive until now. As realization among market participants of these advantages increases, F increases, putting upward pressure on P.

-     **T** is the consensus expectation among market participants of how long it will take for useful innovations to come to market. As the expectation of their arrival decreases from some distant future to relatively soon, T decreases, putting upward pressure on P. Also, as naysayers' and gadflies' expectations decrease from never to maybe one day, the market for virtual currency generally and bitcoins specifically should expand, putting additional upward pressure on P.

- r includes an underlying risk-adjusted return, that includes risk premiums associated with technical difficulties and security issues related to the holding of bitcoins, regulatory uncertainty, market uncertainty related to the widespread adoption of the services listed in the point above, price volatility caused by discoordinated changes in expectations about the three independent variables, etc. As uncertainty decreases, r decreases, putting upward pressure on P.

Granted, every point here includes an implicit "...and vice versa," but it has been common practice among Bitcoin early adopters to draw analogies between Bitcoin in the mid-2010s and the Internet in the mid-1990s, when the answering machine, the fax, and the beeper were the cutting edge of communication technology, and Marc Andreessen had only recently developed the first popular web browser as a student at the University of Illinois (NCSA 2013).

At the time of this writing, Bitcoin had begun to attract the kind of investor attention that we saw with Dot-Com startups in the mid-1990s, and mainstream media pundits had begun to move beyond the *Sturm und Drang* over its nefarious uses, reminiscent of the *Time* magazine cover of 3 July 1995 with the caricature of a shocked child basking in the blue glow of a computer monitor, above the headline, "Cyberporn – Exclusive: A New Study Shows How Pervasive and Wild It Really Is. Can We Protect Our Kids—And Free Speech?" (*Time* 1995). Whether Bitcoin specifically or virtual currency generally will spawn the next generation's 'Apple', 'Facebook', 'Google', etc. remains to be seen, but if it does, then this could lead the kind of positive feedback suggested above.

2.D. Drawbacks of Bitcoin

As there is a cost for every benefit, Barber, et al. (2012), Eskandari, et al. (2015), European Central Bank (2012), Grinberg (2011), Kroll, et al. (2013), Lo & Wang (2014), Velde (2013), Young (2015), and many others point out that Bitcoin has significant drawbacks. This is not intended to be an exhaustive list, and these are known issues that provide entrepreneurial incentive to resolve them:

2.D.i. Understanding and Acceptance

Like the Internet a quarter-century ago, Bitcoin is new and unknown to the vast majority. For it to catch on beyond a relatively small community of tech-savvy early adopters and persons excluded from the mainstream banking system, more individuals need to understand what Bitcoin is and to accept that such a system can live up to its promise. This is facilitated in part by enthusiasts' promoting it to their friends and families and rebutting the criticisms of naysayers and gadflies. While many critiques of Bitcoin are legitimate, one sometimes sees accounts in the popular media and comments on online discussion boards that seem to be willful misrepresentations. As with technological innovations in the past, understanding and acceptance will grow, if entrepreneurs develop solutions based on Bitcoin to problems that *status quo* institutions and technologies have not solved.

2.D.ii. Reputation

Rooney (2011) in a review of work by Genevieve Bell, Director of Intel Corporation's Interaction and Experience Research, notes that for a new technology to induce moral panic, it must change one's relationship with time, space, and other people. The ability to transfer any amount of value across national borders that clear in less than an hour to anyone in the world who has Internet access—including the vast majority of individuals globally who do not have bank accounts but have mobile phones—fulfills all three requirements. Add to this the fact that those who are excluded from or marginalized by *status quo* institutions have greater incentives than those who are well served by them to embrace new technologies that do not exclude or marginalize them. It is seemingly inevitable that many innovations gain initially unsavory reputations, as the unsavory tend to be among the first to embrace them. In time, individuals grow accustomed to innovations, and future generations no longer see them as disruptive, but rather as things that always have been there.

2.D.iii. Trust

Related to understanding, acceptance, and reputation is trust. For Bitcoin to find a market among the general population, users will have to trust claims about its security and stability. Innovators and early adopters are quick to point out that Bitcoin's source code is free and open, meaning that anyone can make a copy of it and analyze it for software bugs and security backdoors. One also could inspect the mechanical integrity of a jet airplane before flying in it, but very few are qualified to recognize what they are looking at. Ultimately, one falls back on logical fallacies including appeals to authority (Executives at money center banks and stock exchanges are experimenting with Bitcoin.), appeals to popular sentiment (The user base is doubling every three months.), and appeals to tradition (Bitcoin has been running nonstop since January 2009 and has never been hacked.).

2.D.iv. Killer App Use Cases

Rogers (1962) describes a process of the diffusion of innovation that is often referred to at the Technology Adoption Life Cycle. It begins with Innovators and Early Adopters who embrace new technology eagerly, and a Mainstream comprised of the majority of the population who embrace new technology only after it has proven its practical utility. Moore (1999) expands on this by positing a 'chasm' between the former to the latter, into which most innovations fall and are forgotten. Entrepreneurs develop so-called 'killer apps' for some few innovations that make them palatable to the majority. As of the time of the drafting of this manuscript, Bitcoin's killer app is not yet manifest.

2.D.v. Transaction Confirmation Speed

In the early days of television in the 1950s, a situation comedy like *The Honeymooners* was filmed on a sound stage before a single camera. The productions were essentially theater plays that were recorded for broadcast. Early use cases for Bitcoin similarly have been 'doing the same thing better'. Initial attempts to promote adoption have tended to focus on getting as many storefront merchants as possible within a geographical area to accept bitcoins in payment for face-to-face transactions. Considering that blocks are generated approximately once every ten minutes, and initial notifications of unverified transactions can take several minutes to transmit over the Bitcoin network from the sender to the receiver—even though they are standing in front of each other, smartphones in hand—a common complaint is that Bitcoin is less efficient than cash, debit cards, and credit cards. As of the time of the drafting of this manuscript several projects have been announced to remedy this. Whether any will succeed is unknowable. Although clearing an international payment of any size within less than an hour and for a fee measured in pennies is vastly faster than using a money transmitter or bank wire, this argument is of little use for someone who wants to pay for a cup of coffee or a taxi ride with Bitcoin.

2.D.vi. Regulatory Uncertainty

Related to understanding and acceptance among potential users is understanding and acceptance among legislators, regulators, and law enforcement officials. A result of this lack of understanding and acceptance can be moral panic (Rooney 2011). Bitcoin users often have operated in legal gray areas, and potential users have remained potential users. Government officials in several OECD member states have begun embracing Bitcoin with varying degrees of enthusiasm, and the general trend has been toward acceptance. Nonetheless, the financial landscape after the passage of the USA PATRIOT Act in 2001 has become a legal minefield. Those who prefer to ask permission to act, rather than to ask for forgiveness after-the-fact, might be deterred from embracing Bitcoin just yet.

2.D.vii. Consumer Protection

When a Bitcoin user sends bitcoins to a receiver's address, the only way to get them back is to ask the receiver for a refund. As a global network that runs on millions of volunteers' computers, Bitcoin has no corporate headquarters and no executives to sue or prosecute. In this way, bitcoins behave like cash. If the sender and receiver are not face-to-face, the transaction begins to feel more like a debit card or credit card transaction, since one cannot transfer cash over the Internet. For example, if one buys a sandwich with cash, and the seller refuses to deliver, the buyer is able to address the dispute in person. When purchasing something online, this is not possible. Many potential users find this transaction irreversibility unsettling, because they can initiate chargebacks when they pay with their credit cards. Pointing out to a potential Bitcoin user that the costs of the vast majority of fraudulent credit card transactions are borne by merchants, and not consumers, is often unconvincing. Possible technological and institutional solutions exist here, although they are not used widely at the time of the drafting of this manuscript.

2.D.viii. Theft and Loss

A subset of consumer protection is theft and loss. Once one's bitcoins are gone, they stay gone. In this way, they are similar to cash and other property. The only way to get around this is to entrust one's bitcoins with someone for safekeeping, which many believe obviates Bitcoin's utility, except perhaps as a speculative store of value, like a conventional investment through a brokerage account.

2.D.ix. Key Management

At the time of the drafting of this manuscript, users have two general methods for holding bitcoins. One is with an online service, and the other is by running software on one's mobile phone or computer (wallet). With online services, one generally does not have access to the private encryption keys that secure one's bitcoins, putting one at risk of losing them if the service provider fails or gets hacked. With wallet software running on one's own equipment, one must take great care not to delete or corrupt one's encryption keys or lose them to malicious software. The state of the art of Bitcoin wallet software is somewhat analogous to how users accessed the Internet from their homes in the early 1990s. This involved the configuring of unintuitive software with arcane commands that only the technically inclined were likely to attempt. Over time, managing one's bitcoins securely should improve as programmers and entrepreneurs address this concern, in similar fashion to how connecting a computer to the Internet today is often automatic, unlike twenty-five years ago. An important difference here is that twenty-five years ago one might have been frustrated trying to configure a modem, and today one risks losing potentially substantial amounts of value.

2.D.x. Price Manipulation, Run-Ups & Corrections

From early 2010, when Bitcoin trading began, through the middle of 2015, the USD price of a bitcoin experienced four major run-ups and subsequent corrections. The first was in 2011, when it went from 75¢ in April to more than $30 in June, crashing to less than $2 in November and finally settling in around $5. The second was in 2012, when it began rising in June to a high a bit above $15 in August, crashing to approximately $7.50 and finally settling in around $13.50 beginning in December. The third was in early 2013, when it went from $13.50 to more than $260, crashing to approximately $50 and settling in around $130. The fourth was in late 2013, when it went from $130 to more than $1,200, crashing to approximately $450 in early 2014, settling in around $900 for some months, and then drifting downward to a low of approximately $150 and eventually settling in between $200 and $300 by mid-2015. While an increase from 75¢ to $300 over 4.25 years yields an average return of 300% per year, such a roller coaster ride is enough to put many off. During the same period, the price has experienced many smaller bursts of irrational exuberance and irrational malaise, suggesting the possibility of price manipulation by pump-&-dump gangs. Beginning in early 2015, the price seemed to stabilize somewhat, as institutional investor interest began to emerge. It also seemed to respond to macroeconomic events in China and Greece, *contra* Ciaian, Rajcaniova & Kancs (2014), although it is too early to tell for certain. This remains an open question for future research.

2.D.xi. Deflation

Some economists doubt Bitcoin's long-term viability as a medium of exchange, because of the ultimate cap on its circulation of 21 million bitcoins, even though each is divisible into 100 million satoshis. The belief is that the fixed quantity will lead to users' unwillingness to part with bitcoins of ever-increasing value in exchange for goods and services in the present. This is based on Keynesian analysis that is beyond the scope of this paper to rebut, other than to note that the modern experiment with inflationary fiat dates back only to the early 1970s. For millennia before that, economies worldwide based on gold or silver as money faced cycles of inflation and deflation, and money continued to perform all of its historical functions throughout. Whether these modern concerns are correct remains an open question for future research.

3. Cryptocurrency Wallets & Storage

In order to use a particular cryptocurrency, one needs to have access to a software application known as a 'wallet' that is specific to that cryptocurrency. However one acquires cryptocurrency, as discussed below, if one holds it, rather than immediately spend it onward or sell it for conventional money, then one must be able to hold it.

3.A. Wallets

In general, one has five options for holding cryptocurrency:

i.   One can install a **software wallet** onto one's own personal computer or smartphone. One can find a list of the most popular cryptocurrencies at Coin Market Cap[26]. (Some wallets work with multiple cryptocurrencies.) One can find links to wallets that work with Bitcoin, the original and most popular decentralized cryptocurrency, at Bitcoin.com[27].

ii.   One can use a physical device that stores cryptocurrency private keys, known as a **hardware wallet**.

iii.   One can use a **service that hosts and manages wallets** on behalf of cryptocurrency users (e.g., Blockchain.info, Xapo).

iv.   One can leave one's cryptocurrency in a cryptocurrency **exchange account** (e.g., Bitstamp, Coinbase, Kraken, Gemini).

v.   One can keep one's cryptocurrency in '**cold storage**', on some medium that is isolated from the Internet.

Each option strikes a different balance between convenience and security.

---

26 https://coinmarketcap.com
27 https://www.bitcoin.com/choose-your-wallet

3.A.i. Software Wallet

If the amount cryptocurrency that one holds is relatively small, then one might maintain control over it (i.e., 'keep it') in a software wallet on one's computer or smartphone. One advantage of installing a wallet on one's own computer or smartphone is that one has direct control over the private keys that define one's cryptocurrency holdings. One disadvantage is that, if one's hard drive crashes, and one has not made secure backups of one's private keys, then one's cryptocurrency is irrecoverable and lost forever.

3.A.ii. Hardware Wallet

If one is security-conscious and technically adept, and one trusts the integrity of the producer of a hardware wallet—a physical electronic device (typically a USB drive or dongle) that stores one's cryptocurrency private keys separate from one's computer (e.g., Ledger Nano, Trezor, KeepKey)—then one can exercise greater security than that enabled by a software wallet, but with the inconvenience of having to keep track of an additional physical device that is small enough to become misplaced easily.

An added advantage of hardware wallets, for those who engage in large-value transactions with trusted counterparties who value privacy very highly, is that one can deposit arbitrarily large amounts of cryptocurrency into a cryptocurrency address that is controlled by a hardware wallet and physically hand the hardware wallet to one's counterparty, as a kind of 21st Century proverbial 'suitcase full of cash' or the proverbial 'bus station locker key'. Similarly, one can maintain a collection of hardware wallets, in order to segregate one's holdings, such that one might use one's hardware wallet for one specific category of transactions, another for a completely different category, etc. In this way, one might keep one wallet for transactions that face the public, as one does with one's bank account; another wallet for sensitive or private transactions, like gambling or other stigmatized activity; yet another for long-term safekeeping; etc. To the outside world, each of these wallets would appear to be an independent 'person', similar to the way that one would maintain separate bank accounts for a group of companies, even if the same natural person were the sole shareholder of all of those companies.

3.A.iii. Hosted Wallet

If one wants the convenience of a software wallet, but without the risk of one's hard drive crashing, then one might use a service that hosts one's wallet[28]. Some wallet hosting services encrypt the users' cryptocurrency private keys, so that the operators of those services cannot access them. With this kind of service, the user has a password—or a string of random words ('passphrase')—that serves as a key to decrypt the private key that the service hosts. Because the host does not know the password or passphrase, its operators cannot recover the user's cryptocurrency, if the user forgets the password or passphrase. Nonetheless, if the wallet host shut down, then the user would lost access to the user's cryptocurrency.

One advantage of using a wallet hosted by a third party is that one does not have to secure backups of one's private keys. One disadvantage is that one does not have access to one's private keys, and must trust the wallet host's operators to ensure that the users have access to their accounts with the wallet host. If the wallet host encrypts the private keys, the users have the double disadvantage of having to trust the wallet host and to secure their passwords or passphrases with the same due care that they would use to secure their private keys, if they used wallets running on their own computers or smartphones.

---

28 The operators of systems like Blockchain.info and Xapo, *inter alia*, commonly refer to the wallets that they manage on behalf of their users as 'wallets', although the relationship between the user and the wallet host more closely fits the conventional description of an *account*, in which the wallet host holds the cryptocurrency on behalf of the user.

3.A.iv. Exchange Account

If one plans to trade cryptocurrency actively, then one might leave it in one's account with an exchange; e.g., Bitstamp, GDAX, Gemini, Kraken, etc. (discussed in detail in §4.). While doing so exposes one to the risk of one's account or the exchange's being hacked, it grants one the convenience of being able to trade speculatively.

3.A.v. Cold Storage

If the amount of cryptocurrency that one holds is large, and one does not plan to transact or trade actively with it—or even hide it—then one might keep it in 'cold storage', meaning some medium that is not connected to the Internet. This can be a) a 'hardware wallet' that one rarely uses, b) a trusted party that provides cold storage services (e.g., Xapo Vault), or even c) in the form of private cryptocurrency keys printed on a piece of paper that is stored in a secure location.[29]

Each of these has its positives and negatives. If the firmware on a hardware wallet is corrupted or the physical medium is damaged, then the bitcoins can be lost forever. Similarly, if one entrusts one's bitcoins to a third party, then one risks the incompetence, negligence, or malice of that party, or statutory or regulatory action against that party.

 Above all, one must remember that—a) although the selling of cryptocurrency as a business is regulated at the federal, state, and sometimes local level; b) tax authorities have issued guidelines on the taxation of cryptocurrency transactions and capital gains; and c) the use of cryptocurrency in transactions involving criminal activity is proscribed—no equivalent of Federal Deposit Insurance Corporation (FDIC) or Securities

---

29 See "The Wealthy Are Hoarding $10 Billion of Bitcoin in Bunkers" for a description of the service that Xapo Vault provides to its customers. https://www.bloomberg.com/news/articles/2018-05-09/bunkers-for-the-wealthy-are-said-to-hoard-10-billion-of-bitcoin

Investor Protection Corporation (SIPC) insurance exists for cryptocurrency exchanges (discussed in detail in §4.) as of the date of this report.

On the one hand, the cryptocurrency holder bears full responsibility for avoiding the risk of mishandling the cryptocurrency under his control. This is particularly critical, if one is holding others' bitcoins, e.g., if one acts as an escrow agent, family office manager, trustee, or other fiduciary, whether formally or informally. On the other hand, if one is technically adept, or if one has access to acquaintances or subordinates who are technically adept, and one is willing to bear the risk of that responsibility, then one can use cryptocurrency to obfuscate one's transactions and hide one's assets.

3.B. Spending & Receiving Cryptocurrency

Bitcoin addresses are strings of numbers and letters that are approximately 33 characters long and begin with a 1 or a 3. One can see how many bitcoins have been received by and spent from any Bitcoin address by searching for that address with a *blockchain explorer*.[30]

Once one has created a cryptocurrency address, one can receive cryptocurrency into it. One can create as many cryptocurrency addresses as one wants. Indeed, some active users create a new address for each transaction, in order to obscure their cryptocurrency activity. This is because distributed cryptocurrency systems are possible, only because everyone—even non-users, including investigators, government officials, law enforcement professionals, etc.—can audit the entire blockchain that records every transaction that has taken place within the system since its inception. By creating a new address for each transaction, tracing transactions is more difficult than if one consistently used the same address all the time.

---

30 One of the most popular and trusted blockchain explorers is at: https://blockchain.info/address/

3.C. Accounts versus Private Keys

From the perspective of *economic theory* and of *daily practice*, if one does not possess a copy of a Bitcoin address's private key, then one does not have direct control over the bitcoins held in that address, and one practically does not own them. One might have *legal title* to those bitcoins, but one's exercise of one's property rights under that title is contingent upon one's ability to ensure that the party that holds the private key—*de facto* an *implied trustee*—obeys one's instructions unquestioningly. While offering opinions concerning the *legal* relationships among presumed *beneficiaries*, *grantors*, and *trustees* is not within the forensic economist's sandbox, describing the *economic* relationships among them is. His role is to alert the legal experts to these issues, and to leave it to the decider of fact to render the final legal judgment.

These issues are particularly important, when one holds one's bitcoins at an exchange account or in a 'hosted wallet'. In a criminal case involving cryptocurrency, the difference between innocence and guilt—the difference between prison time or losing one's citizenship and freedom—can hinge on whether one held *bitcoins*—or units of some other cryptocurrency—or if one held *a claim on bitcoins*. In terms of the discussion in §2.A.ii., did one hold *bitcoins per se*, or did one hold a metaphorical 'meal ticket', 'coat check', or 'valet ticket'?

The relevant analogy here is to a depositor's relationship with his bank or stock broker. When one deposits money into one's bank account or shares into one's stock brokerage account, one does not inventory the serial numbers of the paper USD notes or stock certificates that one transfers to the bank or to the brokerage. Likewise, one does not confirm that the serial numbers of the paper USD notes that one withdraws from an ATM or stock certificates that one sells match those deposited.

Similarly, when one deposits bitcoins—or units of some other cryptocurrency—into an exchange account or into a hosted wallet, one generally deposits them into a pooled account—analogous to a bank vault—and one **receives credit**

within the accounting software of the exchange or hosted wallet provider. Even if the exchange or hosted wallet operator creates a unique Bitcoin address for each user, the user's generally do not have access to the private keys associated with their holdings.

When one withdraws bitcoins from an exchange or hosted wallet provider, one sends a request to the operator. The operator can comply with the request, or it can deny the request for any reason, including court order, the depositor's actual or alleged violation of the service's KYC or AML policies, incompetence or negligence, lack of sufficient funds to fulfill the request, or capriciousness, in the same way that a bank or stock brokerage can freeze a customer's account.

3.C.i. Filing and Disclosure Requirements

Internal Revenue Service (IRS) and Financial Crimes Enforcement Network (FinCEN) regulations require a person who is under US jurisdiction to file an annual disclosure of foreign financial accounts under the Foreign Account Tax Compliance Act (FATCA) of 2010 and a Report of Foreign Bank and Financial Accounts (FBAR) under the Bank Secrecy Act of 1970. The two relevant forms that must be filed are IRS Form 8938 (FATCA) and FinCEN Form 114 (FBAR).

Determining whether the terms of service of an account with a cryptocurrency exchange or hosted wallet service based outside the USA could qualify as a "contract that has as an issuer or counterparty that is other than a U.S. person"—and thus is subject to FATCA declarations and FBARs[31]—is a legal question and outside the forensic economist's sandbox. However, many lawyers are not aware of these requirements, and it is incumbent on the forensic economist to bring them to his client's attention, when potentially relevant to the case.

---

31 See "Do I need to file Form 8938, 'Statement of Specified Foreign Financial Assets'?" at https://www.irs.gov/businesses/corporations/do-i-need-to-file-form-8938-statement-of-specified-foreign-financial-assets.

3.C.i.a. Thresholds

The threshold for IRS Form 8938 is an aggregate value of all foreign accounts greater than $50,000 on the last day of the tax year or $75,000 at any time during the year for a single tax-payer, and an aggregate value greater than $100,000 on the last day of the tax year or $150,000 at any time during the year for a married couple.

The threshold for FinCEN Form 114 is an aggregate value of all foreign accounts greater than $10,000 at any time during the calendar year.

These values thresholds are as of the time that this was written. The forensic economist should verify the current thresholds in force when he is drafting a report that cites them.

3.C.i.b. Penalties

Penalties for failure to file IRS Form 8938 are up to $10,000 for failure to disclose and an additional $10,000 for each 30 days of non-filing after IRS notice of a failure to disclose, for a potential maximum penalty of $60,000. Criminal penalties may also apply.

Failure to file FinCEN Form 114 are, if non-willful, up to $10,000; if willful, up to the greater of $100,000 or 50 percent of account balances. Criminal penalties may also apply.[32]

---

32 See "Comparison of Form 8938 and FBAR Requirements" at https://www.irs.gov/businesses/comparison-of-form-8938-and-fbar-re-quirements.

4. Bitcoin Prices

Most cryptocurrency systems fix either the number of tokens in circulation or the rate at which new tokens are released into circulation, so that the number in circulation in the future is predictable. As of the date of this report, approximately 17 million bitcoins have been released into circulation. The maximum number of bitcoins coded into the Bitcoin software is approximately 21 million. Current estimates are that the last bitcoin is expected to be released into circulation in 2140.

Initially in January 2009, bitcoins had no economic value. The user base was very small, and anyone who wanted bitcoins could run the Bitcoin software on his computer and earn newly issued bitcoins in exchange for contributing computing power to the network. As more users joined the network, it became increasingly difficult to earn bitcoins, and some users began to find it less costly to buy them from those who already held them. The market value of all cryptocurrencies including Bitcoin, taken together as an industry, has grown from nonexistent prior to January 2009 to more than $150 billion at the end of 2018 (Coin Market Cap 2018). If Bitcoin were a firm, its market capitalization would be among the top twenty S&P 500 firms (Standard & Poors 2018).

The first recorded arm's-length transaction involving Bitcoin was conducted 22 May 2010, when Laszlo Hanyecz of Jacksonville, Florida, agreed during a conversation over Internet Relay Chat to pay anyone 10,000 bitcoins for two deluxe pizzas—worth approximately $25—to be delivered to his home that evening. Jeremy Sturdivant agreed and placed the order from his home near Santa Cruz, California. After the pizzas arrived, Hanyecz transferred 10,000 bitcoins to Sturdivant, which he confirmed receiving. This established a price per bitcoin of approximately 0.25¢[33] (Bitcoin Who's Who 2016).

---

33 In Dec. 2017, when the bitcoin price peaked at approx. $19,000, 10,000 bitcoins were worth approx. $190 million. By Dec. 2018, the price had fallen to approx. $4,000, for a value of approx. $40 million.

4.A. Purchase and Sale of Cryptocurrency

The first Bitcoin exchanges opened in 2010, enabling buyers and sellers to trade bitcoins as if they were publicly traded shares. Since then, the price of 1.00 bitcoin has approximately doubled each year on average, punctuated by run-ups and subsequent corrections back down to approximately the previous run-up's high. (See Figure 8.)

The two most common ways to acquire bitcoins are 1) in payment for goods and services and 2) buying them, either from an acquaintance, from an ad hoc counterparty that one has met via a matchmaking service like LocalBitcoins, from an exchange like Coinbase, Kraken, or Gemini for retail quantities, or through a broker via a service like Circle Trade for institutional quantities.

An increasing number of merchants worldwide accept bitcoins, particularly in areas where the banking systems are expensive, inefficient, or inconvenient.

When an individual purchases Bitcoin using a service like a cryptocurrency exchange (Exchange), he becomes a member (Member) of that Exchange by creating an account. The Member transfers money from his bank account to the Exchange's bank account, thereby creating a transaction record both at the Member's and at the Exchange's bank. Alternatively, the Member transfers cryptocurrency from his cryptocurrency software wallet (see §3.) or from another exchange to a cryptocurrency address controlled by the Exchange, thereby creating a transaction record in both the Member's (or other exchange's) and the Exchange's Wallets, which reflect information within the transaction ledger of the cryptocurrency system (in the case of Bitcoin and its spun off 'forks', on the relevant blockchain).

Exchanges are custodians of Members' money and cryptocurrency. While the Member's money is in the Exchange's custody, the Member does not have direct control over it; the Member has an account with the Exchange, and all records of transactions that take place within the Exchange are kept by the Exchange's operators. Once money has been credited to the Member's account with the Exchange, the Member can buy Bitcoin—or other cryptocurrencies—from other Members who have accounts with the Exchange. When the Member buys Bitcoin from other Members, the records of the Member's money and Bitcoin holdings are maintained and controlled by the Exchange's operators. Once cryptocurrency has been credited to the Member's account with the Exchange, the Member can sell Bitcoin—or other cryptocurrencies—to other Members who have accounts with the Exchange. When the Member sells Bitcoin to other Members, the records of the Member's money and Bitcoin holdings are maintained and controlled by the Exchange's operators.

When a Member removes his cryptocurrency from the Exchange, the Member instructs the Exchange to transfer the Bitcoin from the Exchange's control to a Bitcoin address that either is under the control of the Member, is under the control of another Exchange or a merchant that accepts Bitcoin, or is under the control of any other third party. The Exchange's operator is able to deny a Member's transfer request in response to a court order; in compliance with an instruction from a law enforcement or regulatory agency; due to technical failures, including hacking, denial-of-service attack, war, natural disaster, act of God, etc.; based on the Exchange's Terms of Service; through incompetence, negligence, or malice; or some combination of these causes.

4.A.i. US Cryptocurrency Exchanges

As discussed above, buying and selling cryptocurrencies on an exchange involves setting up an account and tying it with one's bank account. One must provide enough personal information to fulfill the exchange's Know Your Customer (KYC) and Anti-Money Laundering (AML) requirements. In order to trade in large quantities, it is necessary to provide even more identifying information: proof of residence, photo identification, etc.

The most reputable retail exchanges in the USA are:

Coinbase:	https://coinbase.com/
GDAX:	https://gdax.com/
Gemini:	https://gemini.com/
Kraken:	https://kraken.com/

Institutional investors and accredited investors can use Gemini and Circle Trade (https://circletrade.com).

4.A.ii. Broker versus Dealer

A search of the websites of the US Attorney's Office and the SEC reveals that many criminal prosecutions involving cryptocurrency involve accusations of operating as an unlicensed money transmitter. At the state level, it can be crucial to distinguish between buying and selling cryptocurrencies out of one's own inventory as a *dealer* or matching buyers and sellers as a *broker*. In some states, operating as a dealer is not regulated, although in most states operating as a broker is regulated.

The legal determination of whether the defendant in a money transmittal case is operating as a dealer or broker—and if this distinction is relevant in the jurisdiction where the case is being tried—is outside the forensic economist's sandbox. However, the economic determination is within his sandbox, and he can advise the lawyers who have hired him that this distinction might be relevant to the case.

Because statutes, regulations, and administrative rules related to cryptocurrency specifically and financial services generally often change in response to innovation, there is no point in publishing a comprehensive list here of state, federal, and international positions. The forensic economist who has been hired as an expert on cryptocurrency in a money transmittal case will have to conduct research on this point while preparing his report and testimony.

4.A.iii. Forks Catch Exchanges Flat-Footed

The unexpected viability and popularity of forked coins (discussed in §2.B.iv.) has created an environment ripe for litigation, because of the obsoleting of the design assumptions underlying Exchanges' software and cryptocurrency storage systems. Before Ethereum and Ethereum Classic forked from each other, the inclusion of a new cryptocurrency among those that an Exchange's operator supported for trading was essentially like deciding which lines of goods to carry in a retail shop. If one initially allowed only Bitcoin trading, one could add Litecoin, Ethereum, Monero, Dash, etc., as one chose, based on whatever standards one adopted, however arbitrary those standards might be. If one did not enable trading of a particular cryptocurrency on one's exchange, then no one could claim any economic harm. The decision was wholly at the Exchange's operator's discretion.

Now, it is as if the goods that one carried in one's retail shop could spawn new lines of goods, in the same way that shares in a stock brokerage account can spawn shares in new firms that are spun off from existing firms. However, unlike a retail shop, and perhaps more like a consignment shop, in this analogy, customers would have claims on the goods in one's shop, and one would have to keep track of which customers' goods had spawned new goods. Either that, or one would have to make a case for seizing or discarding customers' spawned goods. To follow the analogy further, it would be as if one allowed customers to sell a particular brand of canned peas on consignment in one's retail shop and found one morning—like a scene from a story in the world of Harry Potter—that the storeroom was full of cans of the same brand, but now not only canned peas, but canned otters' noses, canned sweet pickled garlic, canned pine bark, etc. If different customers had brought in the original canned peas to sell on consignment, then the shopkeeper would have to allocate the 'forked' cans of in proportion to each customer's holdings of canned peas. If the shopkeeper chose not to display the 'forked' otters' noses, pickled

63

garlic, pine bark, and all the rest, then he either would have to transfer them to his customers, sell them and give his customers the proceeds of the sale, keep them, or discard them.

Complicating matters is the fact that, in an Exchange, Members trade cryptocurrencies with each other continuously, and those who automate their trading are active 24 hours per day 365 days per year, with no breaks for weekends or holidays. Thus, the amount of a particular cryptocurrency that one has a claim to can change from one moment to the next, while the amount of the forked coin that one is due is determined at the precise moment of the fork, as in the Bedford Falls/Pottersville example in §2.B.iv.a..

One problem here is that *anyone* can fork a cryptocurrency. One need only make a copy of the software running a currently operating cryptocurrency, change some parameters in the software to make it incompatible with the existing system, convince others to join the new network, *et voilà*: a newly forked cryptocurrency that every holder of the original cryptocurrency has a claim to, in proportion to his holdings in the original cryptocurrency at the moment of the fork. Whether or not this new forked cryptocurrency has market value will remain to be seen, but it will have a ready-made user base, thereby creating an incentive for at lease *some* of the holders to help create a market in it.

This can create a bookkeeping nightmare, if the Exchange's operator did not anticipate the forking or spinning off of viable cryptocurrencies in the design of its accounting software, especially if it pools Members' cryptocurrency into a single reserve account or a fixed number of reserve accounts, and keeps track of Members' holdings exclusively in the exchange's accounting software. Granted, the Exchange's operator could check the state of the accounting software at the moment of the fork, but—if it chooses not to enable trading of the new forked cryptocurrency on its exchange—it then must a) allocate the new forked cryptocurrency in proportion to Members' claims at the moment of the fork, and transfer or sell the correct amount for each Member individually, or b) sit on the new forked cryptocurrency or discard it; in other words, a) give Members either their forked cryptocurrency or its market value in some other form—e.g., USD or the original cryptocurrency—or b) deny Members access to their forked cryptocurrency or its equivalent value. Option b) is most Exchange's operators' default policy, making them ripe targets for litigation.

A common practice among Exchange operators is to decide on a case-by-case basis, which forked cryptocurrencies they will acknowledge, and to ignore the rest. This creates the potential for civil—and perhaps even criminal, conspiracy, class-action, and antitrust—complaints from Members who demand—in the same way that shareholders in a firm that spins off a division or combination of assets as a daughter firm demand shares in the new spun-off firms—their forked cryptocurrencies, including those that are not traded on Exchanges. In the absence of statutes, regulations, or administrative rules that clearly delineate how to proceed here, the forensic economist who is called to testify on cryptocurrency issues of this kind must map the specifics of the case that he is working on to the closest economic and financial analogues and anticipate and be prepared to rebut the opposition's most likely alternative interpretations. As daunting as this might seem to one who is testifying on matters such as these for the first time, it is perversely comforting to know that the likelihood is high that everyone else in the courtroom or at the deposition is flying by the seats of their pants, often bluffing and blustering their way through cross-examination.

Aside from denying their Members access to their property in the form of the new forked cryptocurrency units, Exchange's operators who presume to pick winners *ex ante* expose themselves to further complaints, if they guess incorrectly. This is particularly nettlesome for those Exchange's operators who presume to predict which of the dozens of Bitcoin forks will establish market prices, even if temporarily before collapsing. If, for example, a forked cryptocurrency establishes a market in the immediate aftermath of the fork, and later collapses, and the Exchange's operator dithered before deigning to transfer the forked cryptocurrency units to its Members, it exposes itself to claims of incompetence, negligence, and malice, and and possibly even theft.

When Ethereum and Ethereum Classic forked from each other in July 2016, Exchange's operators' decision to recognize it was relatively straightforward. It was a one-off event, and they were able to treat Ethereum Classic as if it were an *ab ovo* cryptocurrency. Likewise, when Bitcoin Cash forked from Bitcoin in August 2017, it also looked like a one-off event, and played out in roughly similar fashion to how Ethereum Classic played out.

Then, the floodgates opened, and more than two dozen Bitcoin forks occurred, including: Bitcoin Gold, Bitcoin Diamond, UnitedBitcoin, Bitcoin Hot, Super Bitcoin, BitcoinX, Oil Bitcoin, Bitcoin World, Lightning Bitcoin, Bitcoin Stake, BitEthereum, Bitcoin Top, Bitcoin God, Bitcoin File, Bitcoin SegWit2X X11, Bitcoin Uranium, Bitcoin Pizza, Bitcoin All, Bitcoin Cash Plus, Bitcoin Smart, Bitcoin Interest, Quantum Bitcoin, Bitcoin Lite, Bitcoin Ore, Bitcoin Private, Bitcoin Atom, Bitcoin BitVote, Bitcoin BiZero[34]. While it was impossible to know *ex ante* which of these might have any economic value, each became available to Bitcoin holders in proportion to their Bitcoin holdings as of the moment of the fork of each. According to Coin Market Cap[35], many of the forked cryptocurrencies listed above have or had some economic value greater than zero. It would be unsurprising, if many more forks of many different cryptocurrencies that already trade on Exchanges began to circulate and established market values greater than zero.

In light of this turn of events, one question relevant here is whether an Exchange's operational structure was and still is based on the assumption that forks would be transient, and that therefore it would be unnecessary for an Exchange's executives to enable Members' balances to be associated with segre-

---

34 https://iconow.net/list-of-bitcoin-forks/
35 https://coinmarketcap.com

gated/allocated private keys. A related question is whether an Exchange's executives need to redesign their software, policies, and procedures, in order to avoid inevitable disputes in the future of the kind addressed in a case involving forked cryptocurrencies under an Exchange's current operational structure.

Typically, an Exchange's executives publish vague indications of how they might make this determination in the future, but rarely if ever provide objective standards that Members can use to predict which cryptocurrencies will be supported, and at what level they will be supported, on a particular Exchange. This raises a potential question before a court concerning what constitutes acceptable standards for determining the circumstances, under which an Exchange's executives are authorized to deny their customers access to any benefits that accrue to the assets that their customers have entrusted to them. As they operate now, it is typically impossible for a Member of an Exchange to predict which forked coins its executives will release to him and which they will appropriate for the benefit of an Exchange's shareholders or simply leave unclaimed and inaccessible by the Member.

Other questions before a court in a case involving forked cryptocurrencies might include:

1. Because an Exchange's Members do not transfer cryptocurrency units directly to each other within the exchange, instead transacting via database entries in the exchange's accounting software, is a Member's balance within that exchange a) a *financial liability* of an Exchange to the Member that is denominated in cryptocurrency, b) a *derivative asset* that represents a claim on an underlying cryptocurrency asset, or c) a *virtual currency* that is pegged to another cryptocurrency and circulates within the walled garden of an exchange's accounting software?

2. What is the legal relationship between an Exchange and its Members? Is an Exchange balance a) like a warehouse receipt, in which a Member entrusts his cryptocurrency to a fiduciary who holds it in a segregated/allocated account, such that the Member controls the final disposition of his assets and retains any benefits that accrue to those assets while entrusted to the fiduciary, b) like a bank or a stock brokerage account balance, in which the Member deposits his cryptocurrency into a pooled account and forfeits control over the final disposition of his assets but retains any benefits that accrue to those assets, c) or like a vendor receipt, in which the Member purchases/barters a good or service (e.g., virtual currency pegged to another cryptocurrency; see 1.b in the paragraph above) with cryptocurrency, forfeiting control over the disposition of the cryptocurrency paid along with any benefits that accrue to that cryptocurrency subsequent to the completion of the transaction?

3. In light of the fact that an Exchange's executives enable their Members to enjoy some, but not all, of the benefits of the cryptocurrency entrusted to them—e.g., crediting Members with their Bitcoin Cash, but withholding their Bitcoin Atom—what are acceptable standards for determining the circumstances, under which an Exchange's executives are authorized to deny their Members access to any benefits that accrue to the assets that their Members have entrusted to them?

4. Should a court order the executives of an Exchange to reorganize its operational structure, in order to avoid inevitable disputes in the future of the kind addressed in a case involving forked cryptocurrencies under its current operational structure?

A forensic economist, who has been retained by a defendant's attorney representing an Exchange operator, should make his client aware of the foregoing, be prepared to rebut arguments based on the foregoing, and ensure that the Exchange operator provides technical and legal experts to supplement his economic analysis and testimony. A forensic economist, who has been retained by a plaintiff's attorney in a case involving a complaint against an Exchange should be very careful to stay within the sandbox of his expertise and avoid expressing legal opinions—particularly about trust relationships—and avoid getting pulled down a rat hole on technical issues.

4.B. Bitcoin Price Volatility

As discussed above, the supply of bitcoins in circulation is fixed, until the next mining reward is paid approximately every ten minutes, and the growth rate is known and predictable. Therefore, price fluctuations are driven predominantly by shifts in demand, which can be driven by news, noise trading, herding, etc. In spite of its pronounced gyrations, the bitcoin price has approximately doubled every year since May 2010, and even if one restricts one's sample window to the most recent few years, the rate of increase is approximately the same, or even a little bit higher. Granted, past performance is no guarantee of future performance, but in the absence of evidence to the contrary one might expect Bitcoin's popularity to continue to grow, even if not at the same rate that it has maintained for nearly a decade.

From early 2010, when Bitcoin trading began, through the end of 2018, the USD price of a bitcoin experienced five major run-ups and subsequent corrections:

1) **2011** - from 75¢ in April to more than $30 in June, crashing to less than $2 in November and finally settling in around $5.
2) **2012** - from around $5 in June to a bit above $15 in August, crashing to approximately $7.50 and finally settling in around $13.50 beginning in December.
3) **Early 2013** - from $13.50 to more than $260, crashing to approximately $50 and settling in around $130.
4) **Late 2013** - from $130 to more than $1,200, crashing to approximately $450 in early 2014, settling in around $900 for some months, and then drifting downward to a low of approximately $150 and eventually settling in between $200 and $300 by mid-2015.
5) **Late 2017** - from $200-300 in mid-2015 to more than $19,000 in December 2017, crashing to below $4,000 by late 2018.

Recent research finds that each Bitcoin run-up and subsequent correction has followed a similar pattern, and that the growth of the bitcoin price over the long run follows a relatively stable trajectory. (Wheatley, Sornette, Huber, Reppen & Gantner 2018)

Figure 8
Bitcoin Price (log-scale)
15 July 2010 - 31 December 2018

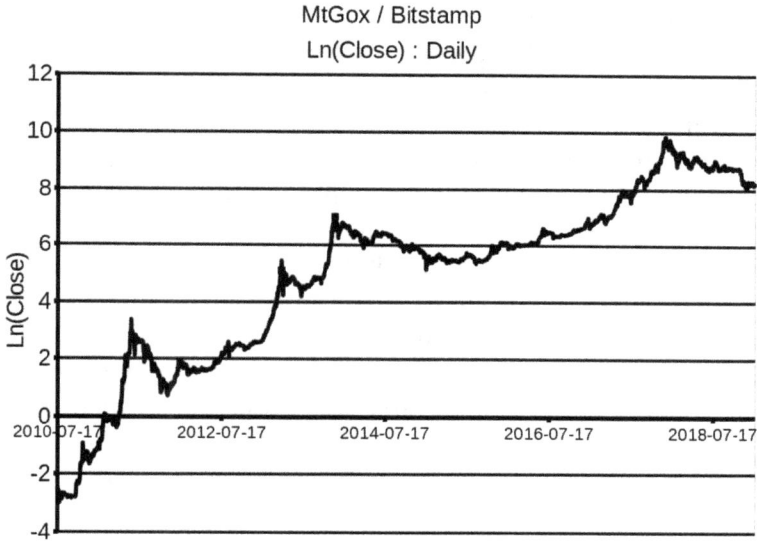

MtGox / Bitstamp
Ln(Close) : Daily

Daily closing prices—15 Jul. 2010 - 31 Dec. 2018—on the MtGOX and Bitstamp exchanges in log-scale, due to the extreme differences in scale from the beginning price of $0.05, through its maximum price to date of $19,377 on 17 December 2017, to its price of approximately $4,000 on 31 December 2018.

4.C. Expectations of Future Value

In some cases, the question of *reasonable expectation of future value* arises. For example, was it reasonable in 2010 to expect the bitcoin price to increase significantly over time? In hindsight, we unequivocally can answer in the affirmative, because this is what, in fact, happened. How about at the time, though, without the benefit of hindsight? It apparently seemed reasonable to those who bought bitcoins, otherwise they would have revealed their preferences by *not* buying bitcoins. In particular, it seemed reasonable to those who bought bitcoins as far back as 2010 and never sold them, even when the price peaked in each of the price run-ups and subsequent corrections described above.

When one buys bitcoins, one does so on the expectation either that they will facilitate transactions that otherwise would be more costly or even impossible using some other medium of exchange, that the bitcoins will increase in value, or both. The bitcoin price at the end of December of each year since 2010 are provided in Table 1.

Table 1
31 December Closing Price of Bitcoin

Date	Price (USD)
2010	0.30
2011	4.75
2012	13.50
2013	730
2014	320
2015	430
2016	965
2017	13,500
2018	4,000

When assessing the reasonableness of an expectation, plan, or action, it is critical to identify the relevant population, against which to compare the action. In the case of a long-term position in Bitcoin, the relevant population is the long-term holders of bitcoins. Those who do not know what Bitcoin is, those who know what Bitcoin is and never have owned bitcoins, those who have owned bitcoins and no longer do, etc. are not relevant comparisons to a long-term Bitcoin holder. The relevant comparison to a long-term Bitcoin holder is other long-term Bitcoin holders.

The question becomes: Do long-term Bitcoin holders hold trivial or significant amounts of value over long periods of time? They do. Table 2 and Table 3 below, list the top 100 Bitcoin addresses that have held bitcoins for more than five years[36] and eight years[37], respectively, and have not spent or sell any of them over that time.

Holding bitcoins through the peak of a run-up indicates that the long-term holder believed that their value was greater than the price at the peak. Otherwise, he would have sold his bitcoins. Instead, the owners of the top 100 Bitcoin addresses listed in Table 2 that had held bitcoins for more than five years, rode out the run-up and subsequent correction.

Table 2 and Table 3 list the top 100 Bitcoin addresses in the first column; the number of bitcoins that each address held at the time that this guide was written; the USD value of those bitcoins at the peak in December 2017; and the value at the end of 2018.

These two tables list only the top 100 Bitcoin addresses. This does not include all of the thousands or perhaps millions of smaller dormant Bitcoin addresses.

Granted, many dormant Bitcoin addresses are essentially lost, because the owner did not keep a backup of the private keys associated with those addresses. However, we can get a sense of approximately what proportion of dormant addresses

---

36 https://bitinfocharts.com/top-100-dormant_5y-bitcoin-addresses.html
37 https://bitinfocharts.com/top-100-dormant_8y-bitcoin-addresses.html

are in fact lost by comparing the values in Table 2, which lists Bitcoin addresses that have been dormant for five years, and in Table 3, which lists Bitcoin addresses that have been dormant for eight years.

At the time of this writing 714,593 bitcoins—worth approximately $12 billion at the peak of the run-up and subsequent correction in December 2017, and approximately $3 billion at the end of 2018—had been dormant for five years. Likewise, 166,662—worth approximately $3 billion at the peak of the run-up and subsequent correction in December 2017, and approximately $670 million at the end of 2018—had been dormant for eight years.

Even if all of the bitcoins that had been dormant for eight years were lost, and we subtracted them from the number of bitcoins that have been dormant for five years, this would leave us with a difference of 547,931 bitcoins, worth approximately $9 billion at the December 2017 peak, and $2 billion at the end of 2018. Even if, hypothetically, half of these bitcoins were lost, this still would mean that the owners of more than $4.5 billion worth of bitcoins rode out the December 2017 peak and the December 2013 peak. If not all of the bitcoins that have been dormant for eight years are lost, then their owners have ridden out all five of the peaks and crashes listed above.

Thus, holders of a substantial number of bitcoins expected that their value would rise, and also that they expected that the value of those bitcoins would exceed the value at the peak of each price run-up and subsequent correction. Thus, one finds support for the contention that it was as reasonably foreseeable for the price of bitcoin to increase hundreds, thousands, and even tens of thousands of percent in value as it is to foresee reasonably the increase in value of any asset, whether real estate, stocks, bonds, options, or whatever.

Table 2: Top 100 Bitcoin Addresses Dormant for 5 Years

Address	BTC	Peak	Today
1FeexV6bAHb8ybZjqQMjJrcCrHGW9sb6uF	79,957	1,364,110,396	319,828,000
1PnMfRF2enSZnR6JSexxBHuQnxG8Vo5FVK	66,452	1,133,707,668	265,808,000
1DiHDQMPFu4p84rkLn6Majj2LCZZZRQUaa	66,236	1,130,022,589	264,944,000
1EBHA1ckUWzNKN7BMfDwGTx6GKEbADUozX	66,234	1,129,988,468	264,936,000
12ib7dApVFvg82TXKycWBNpN8kFyiAN1dr	31,000	528,877,050	124,000,000
12tkqA9xSoowkzoERHMWNKsTey55YEBqkv	28,151	480,271,543	112,604,000
1PeizMg76Cf96nUQrYg8xuoZWLQozU5zGW	19,414	331,213,517	77,656,000
1F34duy2eeMz5mSrvFepVzy7Y1rBsnAyWC	10,771	183,759,184	43,084,000
1f1miYFQWTzdLiCBxtHHnNiW7WAWPUccr	10,009	170,759,044	40,036,000
1KbrSKrT3GeEruTuuYYUSQ35JwKbrAWJYm	10,000	170,605,500	40,000,000
14YK4mzJGo5NKkNnmVJeuEAQftLt795Gec	10,000	170,605,500	40,000,000
12tLs9c9RsALt4ockxa1hB4iTCTSmxj2me	10,000	170,605,500	40,000,000
1P1iThxBH542Gmk1kZNXyji4E4iwpvSbrt	10,000	170,605,500	40,000,000
1BAFWQhH9pNkz3mZDQ1tWrtKkSHVCkc3fV	10,000	170,605,500	40,000,000
1ucXXZQSEf4zny2HRwAQKtVpkLPTUKRtt	10,000	170,605,500	40,000,000
1CPaziTqeEixPoSFtJxu74uDGbpEAotZom	10,000	170,605,500	40,000,000
1HLvaTs3zR3oev9ya7Pzp3GB9Gqfg6XYJT	9,260	157,980,693	37,040,000
167ZWTT8n6s4ya8cGjqNNQjDwDGY31vmHg	8,999	153,527,889	35,996,000
18zuLTKQnLjp987LdxuYvjekYnNAvXif2b	8,021	136,842,671	32,084,000
198aMn6ZYAczwrE5NvNTUMyJ5qkfy4g3Hi	8,000	136,484,400	32,000,000
15Z5YJaaNSxeynvr6uW6jQZLwq3n1Hu6RX	7,941	135,477,827	31,764,000
1DzjE3ANaKLasY2n6e5ToJ4CQCXrvDvwsf	7,000	119,423,850	28,000,000
1FJuzzQFVMbiMGw6JtcXefdD64amy7mSCF	6,999	119,406,789	27,996,000
1AYLzYN7SGu5FQLBTADBzqKm4b6Udt6Bw6	6,512	111,098,301	26,048,000
1JxmKkNK1b3p7r8DDPtnNmGeLZDcgPadJb	6,316	107,754,433	25,264,000
17na83aXEao3jfXJXEN4uuvchgAjg1Mw1S	5,000	85,302,750	20,000,000
19t7WpE7jM3APXAEw8yxhw33Gf1sGn171X	5,000	85,302,750	20,000,000
1ARWCREnmdKyHgNg2c9qih8UzRr4MMQEQS	5,000	85,302,750	20,000,000
1BvNwfxEQwZNRmYQ3eno6e976XyxhCsRXj	4,881	83,272,544	19,524,000
18Hp8j2JMvwtPs1eqNaYEEVvuFpjQJRFVY	4,333	73,923,363	17,332,000
16eb495TbiCRbRbZv4WBdaUvNGxUYJ4jed	4,322	73,735,697	17,288,000

Address	BTC	Peak	Today
1FdjFtrBwf9Jc9fsGN2GtHmG2vs5ZcEuWH	4,154	70,869,524	16,616,000
1GX7i8jG8DD1mG85BNnz7xybVhSmw84Uii	4,109	70,101,799	16,436,000
1ALXLVNj7yKRU2Yki3K3yQGB5TBPof7jyo	4,000	68,242,200	16,000,000
18eY9oWL2mkXCL1VVwPme2NMmAVhX6EfyM	4,000	68,242,200	16,000,000
1LwBdypLh3WPawK1WUqGZXgs4V8neHHqb7	4,000	68,242,200	16,000,000
15MZvKjqeNz4AVz2QrHumQcRJq2JVHjFUz	3,963	67,610,959	15,852,000
1NznGtukoc7Y3bK9VSCJt2B3HfJmJ3oT5Y	3,834	65,410,148	15,336,000
1GMFSWQQQhCQyRNQcac9tDKcvqYCuripVs	3,674	62,680,460	14,696,000
14mPMrRm6TdjqHZhd7aBUbuWt5MYWReukR	3,600	61,417,980	14,400,000
1FvUkW8thcqG6HP7gAvAjcR52fR7CYodBx	3,350	57,152,842	13,400,000
1Gn1GzVa88T1X3fdhejyq6jrZs43T24xW6	3,249	55,429,726	12,996,000
1PTYXwamXXgQoAhDbmUf98rY2Pg1pYXhin	3,233	55,156,758	12,932,000
16oKJMcUZkDbq2tXDr9Fm2HwgBAkJPquyU	3,215	54,849,668	12,860,000
1MQ3UtURCfS96AxC5oZhfWF9eYeRKZUcNr	3,018	51,488,739	12,072,000
19HhmfxGsznL8K7wXjZiFnhqddQucgfZzB	3,000	51,181,650	12,000,000
1JjMoB212ctAiuDvURyWhs813yY4c75cap	3,000	51,181,650	12,000,000
1ArZGb5V24gAgN51FeQknobi6kNyGx739r	3,000	51,181,650	12,000,000
1FDVbVJYKkWPFcJEzCxi99vpKTYxEY3zdj	3,000	51,181,650	12,000,000
1MCG8dJRZpPPZ2Po7jxXLYZnDUSBZKryx3	3,000	51,181,650	12,000,000
15DtovKcGFiAJmyVfbjvCXHyjtyoZhyyj4	3,000	51,181,650	12,000,000
1LQaq7LLoyjdfH3vczuusa17WsRokhsRvG	2,966	50,601,591	11,864,000
19DdkMxutkLGY67REFPLu51imfxG9CUJLD	2,616	44,630,398	10,464,000
1BrSzBwx2RLuppgGziqgF7oMuneHQVhsNc	2,600	44,357,430	10,400,000
17j45BXWrjSDttuurcSQubYLdLescJ7eJH	2,585	44,101,521	10,340,000
1P6brDbYKsQGqRduaTMa1v8hBqJYjig4Qc	2,510	42,821,980	10,040,000
1NT1jtYLNwFXLztD4U4B9sLizdYatirhWW	2,500	42,651,375	10,000,000
12owkvCcMPw5u1M692GbBFmpaMdX3kqXQM	2,500	42,651,375	10,000,000
1NWPS2fWw6FMeJEjg6DMMpYxQyB4TKpVsb	2,343	39,972,868	9,372,000
1Dp1yVTFmgb6oL5WoNVsLsZso4ATMzxD1M	2,288	39,034,538	9,152,000
1F2f54RRbPbj2Svi4Ni9n3RN41W3YwsFPh	2,208	37,669,694	8,832,000
1HjdiADVHew97yM8z4Vqs4iPwMyQHkkuhj	2,200	37,533,210	8,800,000
1JaPNwMXt2AuVkWmkUHbsw78MbGorTfmm2	2,195	37,447,907	8,780,000

Address	BTC	Peak	Today
19dwHtRWNQcAw4gmzmGpYbRbWyj5UqDA3n	2,190	37,362,604	8,760,000
1NB3ZXxs3vfq1hRhuSAZ3zPdQNrXBQB6ZX	2,100	35,827,155	8,400,000
13ACvVSUKeQ57zBtENsJtHManPTJ5sZuRw	2,100	35,827,155	8,400,000
1VLZtmKa95BFrXHeyHEETAivJ22pTEhrT	2,100	35,827,155	8,400,000
16Q5zhKCMbpEkR43K6tgzdkh1mTUfi7SMy	2,100	35,827,155	8,400,000
13DyBwhpDw6152q1drbK2US5S3CdY1mRnU	2,100	35,827,155	8,400,000
1MEKbMf7hPw2scP468unqAXjedZWMM4La4	2,097	35,775,973	8,388,000
1CkdZXJtpbxxX4QAzbRhiFNU3PkcsUsFzw	2,066	35,247,096	8,264,000
1CU33fX35WYJDNxXM5jqawQtVGr32QEGrV	2,050	34,974,127	8,200,000
1BMB272EM8F9RXaFszJ7nxxN8VNjoa3mYu	2,012	34,325,826	8,048,000
16w8WZ8Ub1Whk6SP4cw4op5cgyRVsb77T8	2,000	34,121,100	8,000,000
16y2tVCgnwGM6c3kPPuQDJrSadQqcddUm6	2,000	34,121,100	8,000,000
18KHS8ndbKJ1iEtTxv44Ree3Fs7oCURg83	2,000	34,121,100	8,000,000
1KF8CrJXpAf8EB63a91EHaH1TobnDsRgaK	2,000	34,121,100	8,000,000
1KAhyewvMHSaejzri1jTQD6yvieuHXQ2mP	2,000	34,121,100	8,000,000
1MsgeQMDKCqoPRfKT1tw8gGci6w7fCBtKe	2,000	34,121,100	8,000,000
1NTva3cWy3Xiueh5GPWCa9cz8SePrv6cmT	2,000	34,121,100	8,000,000
13dSnmhFeX3qqbsi4thXXad4ggTh6VCESG	2,000	34,121,100	8,000,000
13RwLs69Y7xPrTM5E2aa9RxiDSyeX6jEyw	2,000	34,121,100	8,000,000
1KxQJau7BkuwHZh7Ugo2yoqKN9WeS51S5C	2,000	34,121,100	8,000,000
1ER3nEZFqXUkKw6AucHnmPxXnQQ8F6SJEG	1,969	33,592,222	7,876,000
14CQ2jCrpsd1eSdC4zWsJZH9LvDr6GrCyo	1,958	33,404,556	7,832,000
1C4H9GDALbjDVnbfTt4a9ML1XbwzLS1rUC	1,908	32,551,529	7,632,000
1FdPpELnjHfwSM4Nvi7LdYS4S4GVGsLUQY	1,804	30,777,232	7,216,000
1Ek9Jj3Z3Bnipe3DnMq2otXG5iNjze66VR	1,799	30,691,929	7,196,000
1NQeyC6ocbxfrzLvYv6VnsJmFZJEu4ufE5	1,783	30,418,960	7,132,000
1DtFKiPdYD2U6XDZGtWK7q8JYVrDKBHBqE	1,766	30,128,931	7,064,000
13FKHnREotr4jrjiSJwPUpecogVT7Rj7bu	1,750	29,855,962	7,000,000
1BVMFfPXJy2TY1x6wm8gow3N5Amw4Etm5h	1,698	28,968,813	6,792,000
1DM1RpNMoFZHFEow9hXYhdTZo2sZRk2TVg	1,666	28,422,876	6,664,000
113324vM6NBar2q72w6iDCdQvPnPQw8Tvw	1,600	27,296,880	6,400,000
1MmcSBpzNYeGgSfVNxXEoBdpmsCxyMUuWW	1,600	27,296,880	6,400,000

Address	BTC	Peak	Today
1JNamaC2PE4woAk1vaoW46e5pxgdQkaKnc	1,592	27,160,395	6,368,000
1JKkR1fSQKY6qUGHdNrD9wzTavButq7AWp	1,526	26,034,399	6,104,000
1J3St3rp3hVsuLGhegbLWicyPL3NRD3YJ7	1,523	25,983,217	6,092,000
1khexYoq6fvMdKAww7dpdZb4WBPKBq2hb	1,518	25,897,914	6,072,000
13oRbW4P5kYaSX4UXAyd5VsSYr8x5hX2Pv	1,500	25,590,825	6,000,000
TOTAL	**714,593**	**$12,191,349,606**	**$2,858,372,000**

Table 3: Top 100 Bitcoin Addresses Dormant for 8 Years

Address	BTC	Peak	Today
12ib7dApVFvg82TXKycWBNpN8kFyiAN1dr	31,000	528,877,050	124,000,000
12tkqA9xSoowkzoERHMWNKsTey55YEBqkv	28,151	480,271,543	112,604,000
1PeizMg76Cf96nUQrYg8xuoZWLQozU5zGW	19,414	331,213,517	77,656,000
1HLvaTs3zR3oev9ya7Pzp3GB9Gqfg6XYJT	9,260	157,980,693	37,040,000
167ZWTT8n6s4ya8cGjqNNQjDwDGY31vmHg	8,999	153,527,889	35,996,000
198aMn6ZYAczwrE5NvNTUMyJ5qkfy4g3Hi	8,000	136,484,400	32,000,000
15Z5YJaaNSxeynvr6uW6jQZLwq3n1Hu6RX	7,941	135,477,827	31,764,000
1FJuzzQFVMbiMGw6JtcXefdD64amy7mSCF	6,999	119,406,789	27,996,000
1ALXLVNj7yKRU2Yki3K3yQGB5TBPof7jyo	4,000	68,242,200	16,000,000
1PTYXwamXXgQoAhDbmUf98rY2Pg1pYXhin	3,233	55,156,758	12,932,000
19DdkMxutkLGY67REFPLu51imfxG9CUJLD	2,616	44,630,398	10,464,000
1HjdiADVHew97yM8z4Vqs4iPwMyQHkkuhj	2,200	37,533,210	8,800,000
13DyBwhpDw6152q1drbK2US5S3CdY1mRnU	2,100	35,827,155	8,400,000
1Ek9Jj3Z3Bnipe3DnMq2otXG5iNjze66VR	1,799	30,691,929	7,196,000
1BVMFfPXJy2TY1x6wm8gow3N5Amw4Etm5h	1,698	28,968,813	6,792,000
1MVLP2kRPNqz8VJUy83LstUoMQzUjgq4Zg	1,201	20,489,720	4,804,000
12vZGymxSfbXmYB6tHYS4oM8kMYYBxDZHx	1,005	17,145,852	4,020,000
14qdBdRTvT4i4QZ6iqRhrBj732x9EpoFQC	1,001	17,077,610	4,004,000
14WWrJMGPo43mAoLXoH53pREAEUAFFW66T	1,000	17,060,550	4,000,000
1EGZ3HcxsG1Vf7YV355CGfN8eqZ5o6wqkY	905	15,439,797	3,620,000
1NS17iag9jJgTHD1VXjvLCEnZuQ3rJED9L	900	15,368,996	3,603,400
137zjnSXZs7Wdhg8zCoAJHz3NPgX8WtPPv	895	15,280,281	3,582,600
16mEzobs4wQPuAMq1C8QSQafcDHvzczVcs	875	14,927,981	3,500,000
1P2ZAuW9nUrFfwgVjfL2SA9sPXSruCfzp8	800	13,648,440	3,200,000
15UkFYLMs5nytwiKWqGgkkVo1fjLFAeJhs	750	12,795,412	3,000,000
16MDruJJaKp7iTXmQ5npFYxyTLfyDEVXyh	600	10,239,742	2,400,800
1ABF1PwZTT4GZCdAPuHF9NSsy78ca2q3wW	600	10,236,330	2,400,000
1FXavuV1rjJicYau2md7pa22Q3P4HEcMLN	600	10,236,330	2,400,000
1Bx8pSCW9Gzyvf1EjkeHJM5DpA3ip4At9h	600	10,236,330	2,400,000
15peGup2HVXUYYdkFmiPLgmg8XX5Ex7jAS	555	9,468,605	2,220,000
15BKWJjL5YWXtaP449WAYqVYZQE1szicTn	550	9,383,302	2,200,000

Address	BTC	Peak	Today
1PxeCXMZBuXHt4CqWWEQ7Kwgdyob9P955L	500	8,530,275	2,000,000
1ELMcoZFxjfdE7WwKKfvm159UnSLr6KmUv	500	8,530,275	2,000,000
1DXnCCDog6rCfGBrUTFw56jkMRN2J7Evbw	500	8,530,275	2,000,000
1CWCrnNi3kWwjvVzsF9LSyJVV9gdp1KYGM	500	8,530,275	2,000,000
1NFx9FETuduha8qJ6ehPYufTqVirB1Jf1D	500	8,530,275	2,000,000
17QBWJCGV4QbQhLbhcS7d9yB2S7DPLf4N9	489	8,344,144	1,956,360
1An4jHFsSz31TjW49vsZtCt2SXKV7oHRWb	405	6,909,522	1,620,000
1G3X24PiZuFBtxAGHnNn1sdT4kyCRQBLjZ	400	6,825,073	1,600,200
1JA4MpuV8MMNYCDTFHdCQeXGyem7mqo4B4	400	6,824,561	1,600,080
19NBWfZniu18DbmneRcaGU3sZCrCvrZrRR	400	6,824,220	1,600,000
12rMpw5HnEvAw3nQqLmRBCQyuktfpa4eVw	400	6,824,220	1,600,000
1MTMKVeNTPu5uvezyCoVkcqp85HKv44GAJ	397	6,773,038	1,588,000
17Sr7mtWpm3rBD4NhwdfdpxqbgU8wPvHhx	353	6,024,080	1,412,400
1JSxDnLYD4XKTQ73N7in2M9XRovw5LANiu	350	5,971,874	1,400,160
1JqPFnGPhHhy54zJKmC1MPiczzgFjCmzE9	340	5,800,587	1,360,000
1H3vfXRb6Lx5cvXCXxfktAyv8cJUaqDRxR	304	5,192,378	1,217,400
1JFs9RUBaqg7rNWGVLv4gZEED7X5GYMw1w	303	5,176,512	1,213,680
12Tzj6dBg1RXss4NyMNvW1T3SkBXNdaKPq	300	5,118,165	1,200,000
1JfKgG3cnupxHb51i9TsZfA3GDc6p3uRqb	300	5,118,165	1,200,000
1ADjwtsvj6QbzCJzeVSoaDiahMK6z5QHPJ	260	4,435,743	1,040,000
1CzJQHjyQshJbXwfLZJdh8pcfLi4bbMfPW	253	4,328,943	1,014,960
19QKDUJtx9n7Vaga6nX1bVHdsnT4Khfyi6	250	4,265,137	1,000,000
1DFhVL3LWnJfi4812VFUBhL1ECWsJdxuVM	250	4,265,137	1,000,000
12CRW6hCS2k1ZUtxShHBbNmL6jUR16Scpm	250	4,265,137	1,000,000
17KTXCuWjc9PT6RJAavaecAyBERCtiraaD	250	4,265,137	1,000,000
15LJefP26GX9CKPQM3idAVdRpQkiw9uFFa	250	4,265,137	1,000,000
1Nbf357fE5mM3Qz97E1aVxqbu2jquJwmDD	250	4,265,137	1,000,000
1Apj1LPGAeEk3i78mXVGqdhpR1rYfuwB83	210	3,582,715	840,000
13GvAdkFeHFGVxTHzcA2rD2e5BD4cGkbBH	200	3,412,621	800,120
1JCeMgVeDzLdxz3G5vRin2ydNxUp6E5yFf	200	3,412,110	800,000
13eA7V7N8vihHCxpHf2CvqUVs7ZnyqJ9ot	200	3,412,110	800,000
1Latnc4VXYzowUNM7opKKTz4Ft4qcJsA1k	200	3,412,110	800,000

Address	BTC	Peak	Today
15QezNwA5ThiPf7wo89TTnfBwny93VQFTp	200	3,412,110	800,000
17ZD3W1v8uXnoQf8vE6i32dZ38tCV7UV1d	200	3,412,110	800,000
132dNJG9EX5DGUuCKvj9npiFcoCbHThLEP	200	3,412,110	800,000
1DTy9z4JvtqYsg44oagVpHqyQpF7ZLLs45	200	3,412,110	800,000
1851awCb56xL78FE7SGWGsdDduuKaZpG3N	200	3,412,110	800,000
1E87cVPLZ938w7vYEA1e9RWSc8mESPA3J5	198	3,377,988	792,000
1BgXk11kqNSbg8Z5KDQuSZhwKqPDse8utP	192	3,278,014	768,560
1LQB1z1r3t1wptDEhMhdw75n9RVeob6w3u	184	3,147,671	738,000
1FaMeHiLHAjGf5dUyVFpKNAc2KSgvXGN8N	166	2,832,904	664,200
1PsW573zho3Wp37zj251JFHc3ydFYDQHUC	155	2,644,385	620,000
16a7dYtHXh5xJ2iMEH9WQuhsUanHFHR7gM	151	2,576,143	604,000
12u6hecWRHEPceLYZ9yhubZvhgKYuLGf1J	151	2,576,143	604,000
16epRTwhrkESCSpqiF6fATB2GZfCnMDDmd	150	2,559,253	600,040
1EQkvhEsewdmDuvNTSgvYafu79F8Q4B3CW	150	2,559,082	600,000
1NBNWXmmMCi9ZSNeC6jwCYVw5gKYpfqUZi	150	2,559,082	600,000
1C16qVq3kZpfjCXAKvrrppqxNmRcnfbFdu	150	2,559,082	600,000
1C3J4uWTdpc3NoJWsCqdpqEjSViyxRXkF	150	2,559,082	600,000
13s48zZJgKPamFmVT4NoZzSJ1iiJ3z7xZh	150	2,559,082	600,000
1BL8TforaKvD1m5bGfZKm69n4T9LT5Quit	150	2,559,082	600,000
1AuudkbLDX3k6CrQ45aAXP4hb1Cr3Ap4hh	150	2,559,082	600,000
15KSPqr6d8hwE2DUbUXGt6yYfHpeSGGBsS	150	2,559,082	600,000
15hMprBNTVjuMss3kpdgh1WTxXcU4Fo1Ts	150	2,559,082	600,000
1G8AdVwkUzQb7ozBEpURXmGT5enf8Mn8Mz	150	2,559,082	600,000
17JixXmjbNiXUz4dg5AXu53n19hwu9zGAV	150	2,559,082	600,000
1JNogxAscT2fQBAanumnLVR8ixG74Qhe3q	150	2,559,082	600,000
183ERa7V3yHrzgrgTjXjundtYBaKFgJYGB	150	2,559,082	600,000
1D5AeWq6eoNt1t9aGJMTSRnWG8Kx8k6KTA	150	2,559,082	600,000
1FzZzMZCVD4NLQMforDVEUKJDwRmmG6iW	150	2,559,082	600,000
1FpcuJR582rb7JYKXrPok1X5hMUAszeBqx	149	2,542,533	596,120
1KL7ZW7R3JReqRhxzK6zKiRVEcXd1nVcA2	144	2,464,055	577,720
19bdnnnKHRzMp6VbPMSxxmpQfEvfJf2P8K	140	2,390,183	560,400
1JUykF46aKocyQMb8TU3xvpLUpotTALFNz	136	2,323,817	544,840

Address	BTC	Peak	Today
1Q5upEUZXRnsrbeZ54rfqGD6Zb996pNarV	133	2,279,460	534,440
14ZKzcF9zF7ZLgAo7H3H1mkiPA6cv9Bwd8	119	2,041,294	478,600
15HoBbCVzxbGVGxxztaP2FAmsCrHHGUp8c	111	1,903,786	446,360
12LTzar3bNDGd26APbWxK3eMArtCVYEhBK	110	1,889,967	443,120
1GaC5UDjTRMvk5kifkb4dkPtDsTB8ubCGP	105	1,793,916	420,600
TOTAL	**166,662**	**$2,843,358,861**	**$666,651,160**

Conclusion

This guide provides economic and technical background for forensic economists who might be called to testify in criminal or civil cases involving cryptocurrency. While a deep understanding of the underlying mathematics, cryptography, and protocols is not necessary, some high-level understanding of the technology that drives Bitcoin is necessary, in order to establish oneself as qualified to testify. Appendix 1 contains a list of resources that can be very helpful, when preparing a forensic economic report.

Since its inception in January 2009, Bitcoin and the cryptocurrency copycats and competitors that it has inspired have grown from a single experiment to run a self-contained, distributed, trustless computer network with no market value into an industry with a market value that rivals most S&P 500 firms. While this still is relatively small, the exponential growth rate of Bitcoin's and other cryptocurrencies' market values over a decade suggests that this value could become significant in the relatively near future.

Unsurprisingly, given that Bitcoin operates outside the international banking system, and that it is impossible to prevent anyone from accessing it, it has attracted some users, who have attracted the attention of regulators, law enforcement investigators, and lawyers. As a result, demand for expert witnesses in cases involving Bitcoin is increasing.

Perhaps the most important thing for a forensic economist new to Bitcoin to bear in mind is that, because Bitcoin is new and innovative, experts, who are qualified to explain to judges and jurors what cryptocurrency is, are very rare at the time that this paper is being written. However, with a bit of study, one can progress from Novice to Expert, compared with the general public, in a relatively short time. From the perspective of the forensic economist, Bitcoin is not proverbial rocket science. Whether one describes it as non-existent buckets of magic sand, a box of poker chips, or a stack of Monopoly™

money that cannot be counterfeited, the end result is that Bit-coin transactions are barter (IRS 2014), which falls within the purview of forensic economics.

For example, cigarettes have been used as money in prisoner of war camps and prisons (Radford 1945). Today, street-level drug dealers in some neighborhoods use bottles of Liquid Tide™ as money (Associated Press 2012), in order to avoid the severe financial crime penalties that can result from transacting through the banking system (Zarate 2013a; 2013b). After cigarettes were banned in US prisons, prisoners began using tins of canned fish as money (Durden 2017). At least one drug dealer in the Netherlands was found accepting Lego™ sets in payment for drugs (*Telegraaf* 2013). A worldwide counterculture of community and alternative currencies is active that exists outside the mainstream economy (Ernstberger2009; Owen 2009; Rahn 1999; Seyfang & Pearson 2000). In all of these environments, a forensic economist is able to estimate economic magnitudes, calculate past and discounted future values, and draft reports based on these analyses that can be replicated. In this sense, transactions involving cryptocurrency are no different.

A Closing Parable

Suppose [that] we have a completely unregulated [blockchain management] system [(BMS)], and an advanced society in which it is economic to carry out all transactions through the accounting system of exchange provided by [this BMS]. The system finds no need for currency or other physical mediums of exchange, and its numeraire has long been a real good, say [gold or central bank fiat]... Suppose now that, for whatever reason, [Satoshi Nakamoto] decides that it would be more aesthetic to replace [gold or central bank fiat] as numeraire with a pure nominal commodity which will be called a [bitcoin,] but which has no physical representation. Although monetary theory has long since passed away, value theory has strengthened with time, and [bitcoin users] realize that the [bitcoin] cannot be established as numeraire by simple decree. It must be a well-defined economic good, that is, the [bitcoin] needs demand and supply functions which can determine its equilibrium value in terms of other goods. Controlling the supply of [bitcoins] is no problem, but creating a demand for them is another matter since [many believe that] they have no intrinsic usefulness. (Fama 1980, 55-56, paraphrase)

Black (1970), Fama (1980), Greenfield & Yeager (1983), and Hall (1982a, 1982b) speculate that the historical functions of money—medium of exchange, store of value, unit of account, and measure of value—can be separated and that each function can be performed by different means. Rahn (1999) predicts that the use of national currencies as mediums of exchange should diminish over time, and that persons will transfer title to assets directly, in order to settle debts and to complete transactions. Cronin (2012) confirms that money is used less over time to clear transactions among banks, as they rely on the canceling of payables against receivables.

Some degree of monetary separation has prevailed at least since the USD became decoupled from gold in the early 1970s. It is possible that national currencies could become Blackian numeraires that continue to serve as units of account and short-term measures of value, and that Bitcoin could assume the role of general medium of exchange and store of value.

How regulators will respond to this remains to be seen. It will be interesting to see if Bitcoin use is regulated using some single existing category—e.g., banking, money transmittal, commodity speculation, currency exchange—or if its use as a medium of exchange will be regulated differently from its use as a store of value, and as whatever other clever and unanticipated uses programmers and entrepreneurs develop.

Following Lessig (1999) regulating Bitcoin according to the rules that apply to the metaphor *du jour* could become as unwieldy as regulating traffic on Interstate highways according to rules written before the invention of the automobile. Bitcoin is a system for keeping track of an intangible asset that exists only for the system to keep track of it, and that does not represent a claim on any other asset or underlying right. Asking if it is a currency, a commodity, a trade secret, a contract, a database, a transaction log, or a bank is like asking if an elephant is a fan, a snake, a tree, a rope, or a wall.

Further Reading

Academic interest in Bitcoin began shortly after the establishment of the first two formal Bitcoin trading forums: Bitcoin Market in February 2010 and the game card trading website, *Magic: The Gathering* Online Exchange (MtGOX) in July 2010, both now closed. One of the earliest contributions was by Grinberg (2011), whose fifty-page paper continues to be a concise and accessible introduction to Bitcoin. In addition to providing explanations of economic and technical issues, he raises several legal questions that are reminiscent of the Easterbrook (1996) and Lessig (1999) debate over The Law of the Horse with regard to *cyberspace*; to wit, whether existing statutes should apply to technological innovations that significantly break from the past—as when automobiles and later airplanes first became popular—or if legislators should enact new statutes written specifically for the innovation.

The choice can have statutory and regulatory repercussions, as Brito & Castillo (2013) point out in the context of Bitcoin. With any new technology, the language often lacks words for new concepts associated with the innovation. This leads us to use existing words in strange, new ways, as our ancestors did when faced with the 'iron road' (railway), 'horseless carriage' (automobile), 'talking picture' (movie), or 'information superhighway' (Internet). (Graef 2010). Many words have very different meanings in different contexts, which can lead to confusion. When we use words to invoke analogies and metaphors, this can lead to our using terms like 'coins' and 'mining' to describe the process by which new bitcoins come into existence, even though there are no coins in Bitcoin. Even if there were, coins are not mined. Ore is mined, ingots are smelted from ore, and coins are minted from ingots. The terms 'coin', 'mining', and 'virtual currency' in the context of Bitcoin are metaphors.

If one refers to Bitcoin metaphorically as a currency, this implies that the statutory and regulatory regime concerned with currencies should apply. One problem with this is that legal codes worldwide generally assume that currencies have centralized issuers that are either governments or central banks within the jurisdictions of those governments. If one refers to Bitcoin metaphorically as a commodity, then this implies a different statutory and regulatory regime that might or might not involve the imposition of value-added taxes, etc. If one refers to Bitcoin metaphorically as a transaction ledger, this implies that someone is holding something somewhere that the transaction ledger keeps records of, suggesting some kind of trust or banking relationship, which implies yet another statutory or regulatory regime.

This situation has created a bit of a cottage industry in analyses of what, precisely, Bitcoin is. For example, Berta & Noonan (2015) argue that units of bitcoin that circulate within the Bitcoin system embody a duality, and that they should be treated as trade secrets, when they are in storage and at rest, and as contracts, when they are in motion between sender and receiver. Raskin (2015) applies existing rules of civil procedure and argues that bitcoins should be treated the same way as tangible property. However, Simovitz (2016) points out that the legal fiction of treating an intangible asset as if it occupied a physical location leads to "a fog of conflicting and arbitrary rules that has clouded enforcement of judgments" (abstract). Luther & Olson (2015, 23) argue that bitcoins can be viewed as 'memory', meaning bookkeeping entries with no intrinsic value that serves as "a publicly observable (and instantaneously updated) record of past transactions that buyers and sellers can consult prior to transacting." Meanwhile, online discussions among Bitcoin users indicate that a plurality, perhaps even a majority, of Bitcoin users want bitcoins to be treated as units of currency alongside the USD, arguing that bitcoins exhibit the classical characteristics of a monetary asset—durable, fungible, divisible, portable, scarce, recognizable, etc. (Barber, Boyen, Shi & Uzun 2012; Meikle 1994; White 1902; Menger 1892)—and that they seem to be a more likely candidate in the 21st Century than, e.g., cigarettes, which have served as a medium of exchange, store of value, and even unit of account and measure of value in some prison settings (Radford 1945).

When regulators and law enforcement officials try to make sense of all this and apply it in practice, it leads to a variant of the story of the Blind Monks and the Elephant. To one regulator, Bitcoin looks like a money transmittal system (FinCEN 2013). To another regulator, a unit of bitcoin looks like a commodity (IRS 2014; CFTC 2014). To yet another regulator, a unit of bitcoin might look like a non-voting capital share of stock in a 'distributed autonomous company' (Larimer 2013). Given that the Bitcoin system performs the function that Fama (1980, 39) defined as "the main function of banks in the transactions industry," to wit: "the maintenance of a system of accounts in which transfers of wealth are carried out with bookkeeping entries," one could argue that the Bitcoin system is a bank, and that bitcoins are deposit liabilities of that bank. To countless Bitcoin supporters, bitcoins look collectively like a community currency (Blanc & Fare 2013; Evans 2014; Owen 2009; Seyfang & Pearson 2000). Whereas, to central bank researchers, a unit of bitcoin looks unlike a currency (European Central Bank 2012; Lo & Wang 2014; Velde 2013). Although, officials at the Canadian Senate have been receptive to Bitcoin (Gerstein & Hervieux-Payette 2015), and US Federal Reserve officials have said, "Virtual currencies bring with them both opportunities and challenges, and they are likely here to stay... Banks need not turn away this business as a class, but they should consider the risks of each individual customer." (Young 2015)

This confusion results from the fact that specific regulators oversee specific subsets of regulated activities. It is understandable that they would side with Easterbrook (1996) and view Bitcoin in terms of a preexisting category over which they have authority (Brito & Castillo 2013), rather side with Lessig (1999) and take a *laissez faire*, hands-off approach—in the post-USA PATRIOT Act, post-Dodd-Frank Act regulatory state (Zarate 2013)—until legislators draft new statutes or create some new regulatory agency to oversee this weird new stuff that defies conventional definitions.

Dealing with this kind of deviation from conventional categories is not unprecedented. We already have grown accustomed to a USD that no longer fits historical definitions of money, in which money [gold] was seen as the 'coat' and paper currency merely as 'claim tickets' (White 1902). Within the context of the conventional banking system, Cronin (2012) revisits key elements of NME—built on the speculative work of Black (1970), Fama (1980), Greenfield & Yeager (1983), and Hall (1982a, 1982b)—that explores issues associated with the separation of monetary functions: chiefly, medium of exchange and unit of account, and less so store of value and measure of value. He concludes that the observed decline in demand for government-issued base money in many jurisdictions—coupled with banks' increasing reliance on the matching and canceling of payables against receivables to clear transactions—provides *prima facie* evidence against theoretical arguments opposing monetary separation.

Prior to the release of eCash (Chaum, Fiat & Naor 1990; Levy 1994) and the development of the Internet as a resource and a venue for mainstream users in the mid-1990s, the hypothetical world that Black (1970) describes bore essentially no resemblance to a world dominated by domestic national currencies and USD-denominated international trade. Black (1970, 40), citing Vickrey (1964; 1955), points out, "[C]urrent monetary theory depends heavily on a rather restricted form of financial institution... [and] other institutional arrangements would make current monetary theory almost completely invalid." He then describes an economy in which the government does not mandate a monopoly national currency that is created by regulated banks, but instead defines a numeraire and allows banks to operate as unregulated portfolio managers. Although their focus is not on the work of Black or Vickrey, Luther & Olson (2015) describe Bitcoin as such an accounting system. It was another decade before Fama (1980), Greenfield & Yeager (1983), and Hall (1982a; 1982b) picked up Black's thread, but this did not develop into a significant research topic among mainstream macroeconomists. However, with the experience over the past quarter-century of platform-specific or in-game currencies like Linden Dollars, Reddit Gold, Facebook Credits, etc., and general-purpose systems like M-Pesa in Kenya that use credit for telephone minutes as money, interest in privately issued payment media is growing (Ernstberger 2009; Gans & Halaburda 2013).

Rahn (1999) notes that national currencies, following the decoupling of the USD from gold in the early 1970s, no longer serve as long-term stores of value or long-term measures of value. Because of institutionalized inflation under floating fiat, a dollar's worth of land, labor, or capital a half-century ago is a very different quantity from a dollar's worth today. Instead of hoarding cash, savers today invest in marketable assets and hold cash only long enough to complete transactions and pay debts. He applies NME to e-currency and predicts that the use of national currencies as mediums of exchange should diminish over time, as financial technology progresses to the point where persons can settle debts and complete transactions by transferring title to assets directly, rather than have the buyer exchange the assets for money and transfer the money to the seller, who then uses it to buy assets. This would leave only unit of account—Black's numeraire—as money's last remaining function as money.

Ciaian, Rajcaniova & Kancs (2014) combine the models of Buchholz, Delaney & Warren (2012), Kistoufek (2013), and van Wijk (2013) to test the determinants of bitcoin prices. They find evidence supporting the hypothesis that bitcoin prices are driven by fundamentals specific the market for bitcoins and investor perceptions of their attractiveness, and that macroeconomic developments have insignificant impact. Even though their value in terms of national currencies is highly volatile at this time, its long-term trend since Bitcoin's inception has been generally upward, making it a potential store of value. Bitcoins already are being used as a medium of exchange worldwide.

That bitcoins do not have 'intrinsic' value is unexceptional, as Luther and Olson (2015), citing Kocherlakota and Wallace (1998) and Kocherlakota (1998a; 1998b; 2002a; 2002b), point out. They argue that bitcoins are a practical demonstration of money as 'memory', in the sense of something intrinsically worthless that is used as a public record-keeping device. Indeed, Bernstein (2008 [1965]) points out that gold is largely useless, and that its highest and best use is sitting on a shelf in a locked vault. It is the ideal monetary asset, specifically because it has very few alternative uses, aside from jewelry and decorating cathedrals.

A detailed analysis of how and why a virtual currency system like Bitcoin works is beyond the scope of this paper. However, Barber, Boyen, Shi & Uzun (2012) and Kroll, Davey & Felton (2013) provide excellent overviews. Readers, who are interested in a deeper understanding how the 'plumbing' works, are directed first to Schneier's (1996) *Applied Cryptography* and then to the seminal work on public key cryptography of Diffie & Hellman (1976), who developed the mathematics underlying public key cryptography; Rivest, Shamir & Adelman (1978), who released the first public key cryptography software, known as RSA; Chaum (1988; 1985; 1982)[38] and Chaum, Fiat & Naor (1990), who describe an untraceable electronic cash system based on Chaum's earlier work, and that he commercialized as eCash through his company DigiCash (Levy 1994). Finally, Satoshi Nakamoto's (2008) pseudonymous self-published White Paper describes his, her, or their proposed Bitcoin system that was released in January 2009 and has been gaining popularity ever since.

---

38 These are only a representative sampling of Chaum's work on e-currency. He maintains an extensive list of his publications at: http://chaum.com/publications/publications.html

References

Akkoyunlu, E. A., K. Ekanadham, and R. V. Huber (1975), "Some Constraints and Tradeoffs in the Design of Network Communications," *ACM SIGOPS Operating Systems Review* 9(5), 67-74. doi: 10.1145/1067629.806523

Antonopoulos, Andreas M. (2015), *Mastering Bitcoin*, North Sebastopol, California: O'Reilly Media.

Associated Press (2012), "The Tide Theft Phenomenon Is So Bad Stores Are Attaching Anti-Theft Tags to Detergent Bottles," *Business Insider* (online, 14 March). http://www.businessinsider.com/tide-theft-drug-dealers-2012-3?op=1

Barber, Simon, Xavier Boyen, Elaine Shi, and Ersin Uzun (2012), "Bitter to Better: How to Make Bitcoin a Better Currency," *Financial Cryptography (FC 2012)* 7397 of Lecture Notes in Computer Science, 399-414. http://robotics.stanford.edu/~xb/fc12/bitcoin.pdf

Bernstein, Peter L. (2008 [1965]), *A Primer on Money, Banking, and Gold*, Hoboken, NJ: John Wiley & Sons.

Berta, Michael A. and Willow W. Noonan (2015), "The Property-Contract Duality of Bitcoin," *Financier Worldwide* June. http://www.financierworldwide.com/the-property-contract-duality-of-bitcoin/

Bitcoin Who's Who (2016), "A Living Currency: An Interview with 'Jercos', Party to First Bitcoin Pizza Transaction," Bitcoin Who's Who [Website] 30 January. http://bitcoinwhoswho.com/blog/2016/01/30/a-living-currency-an-interview-with-jercos-party-to-first-bitcoin-pizza-transaction/

Bitcoin Wiki (2018a), "Category: History," Bitcoin Wiki [Website]. https://en.bitcoin.it/wiki/Category:History

—— (2018b), "Controlled Supply," Bitcoin Wiki [Website]. https://en.bitcoin.it/wiki/Controlled_supply

Bitcoin.com (2017), Website. https://bitcoin.com

Black, Fischer (1986), "Noise," *Journal of Finance* 41(3): 529-543.

—— (1970), "Banking and Interest Rates in a World without Money: The Effects of Uncontrolled Banking." *Journal of Bank Research* 1, 9-20.

Black, Henry Campbell (1910), *A Law Dictionary* 2nd ed. St. Paul, MN: West Publishing. https://thelawdictionary.org/letter/m/page/78/

Blanc, Jérôme and Marie Fare (2013), "Understanding the Role of Governments and Administrations in the Implementation of Community and Complementary Currencies," *Annals of Public and Cooperative Economics*, 63–81.

Brito, Jerry and Andrea Castillo (2016), *Bitcoin: A Primer for Policymakers*, Fairfax, VA: Mercatus Center. http://mercatus.org/publication/bitcoin-primer-policymakers

Buchholz, Martos, Jess Delaney, and Joseph Warren (2012), "Bits and Bets, Information, Price Volatility, and Demand for BitCoin," Working Paper. http://www.bitcointrading.com/pdf/bitsandbets.pdf

CFTC (2014), "CFTC Issues Notice of Temporary Registration as a Swap Execution Facility to TeraExchange, LLC," *Release: PR6698-13* http://www.cftc.gov/PressRoom/PressReleases/pr6698-13

Chaum, David (1989), "Privacy Protected Payments: Unconditional Payer and/or Payee Untraceability," *Smartcard 2000: The Future of IC Cards —Proceedings of the IFIP WG 11.6 International Conference on Smartcard 2000* (David Chaum & I. Schaumuller-Bichl, Eds.), Amsterdam: North Holland, 69-93.

—— (1985), "Security without Identification: Transaction Systems to Make Big Brother Obsolete," *Communications of the ACM* 28(10), 1030-1044. http://chaum.com/publications/Security_Wthout_Identification.html

—— (1983), "Blind Signatures for Untraceable Payments," *Advances in Cryptology Proceedings of Crypto* 82(3), 199–203. http://chaum.com/publications/Chaum-blind-signatures.PDF

Chaum, David, Amos Fiat, and Moni Naor (1990), "Untraceable Electronic Cash," *Proceedings on Advances in Cryptology* (Santa Barbara, California, United States), S. Goldwasser, Ed. New York: SpringerVerlag, 319-327. http://chaum.com/publications/Untraceable_Electronic_Cash.pdf

Ciaian, Pavel, Miroslava Rajcaniova, and d'Artis Kancs (2014), "The Economics of BitCoin Price Formation," Working Paper. http://arxiv.org/pdf/1405.4498.pdf

Chicago Board Options Exchange, 2019, "Overview of S&P 500: Index Components," Website. http://www.cboe.com/products/snp500.aspx

Coin Market Cap (2019), [Website]. http://CoinMarketCap.com

Cronin, David (2012), "The New Monetary Economics Revisited," *Cato Journal* 32(3), 581-594. http://object.cato.org/sites/cato.org/files/serials/files/cato-journal/2012/12/v32n3-7.pdf

Diffie, Whitfield and Martin E. Hellman (1976), "New Directions in Cryptography," *IEEE Transactions on Information Theory* 22(6), 644-654.

Durden, Tyler [pseud.] (2017), "Prisoners Explain Why a Pack of Mackerel Is the Gold Standard of Currencies in America's Prisons," *Zero Hedge* [Website]. http://www.zerohedge.com/news/2017-03-06/prisoners-explain-why-pack-mackerel-gold-standard-currencies-americas-prisons

Dwork, Cynthia; Naor, Moni (1993), "Pricing via Processing, Or, Combatting Junk Mail, Advances in Cryptology," *CRYPTO '92: Lecture Notes in Computer Science No. 740*. Springer: 139–147.

Easterbrook, Frank H. (1996), "Cyberspace and the Law of the Horse," 1996 University of Chicago Legal Forum, 207-216. http://chicagounbound.uchicago.edu/journal_articles/1148/

Economist (2014-01-28), "DAC Attack," *The Economist*. https://www.economist.com/babbage/2014/01/28/dac-attack

Ernstberger, Philip (2009), "Linden Dollar and Virtual Monetary Policy," University of Bayreuth Forschungsstelle für Bankrecht und Bankpolitik Working Paper. http://papers.ssrn.com/sol3/papers.cfm?abstract_id=1339895

Eskandari, Shayan, David Barrera, Elizabeth Stobert, and Jeremy Clark (2015), "A First Look at the Usability of Bitcoin Key Management," Briefing Paper, NDSS Symposium 2015. http://www.internetsociety.org/doc/first-look-usability-bitcoin-key-management

European Central Bank (2012), *Virtual Currency Schemes*, Frankfurt: European Central Bank. http://www.ecb.europa.eu/pub/pdf/other/virtualcurrencyschemes201210 en.pdf

Evans, Charles W. (2015), "The Blind Economists and the Elephant: Bitcoin and Monetary Separation," *Southwestern Journal of Economics* 11(1), 106-121. https://www.s-lideshare.net/CharlesEvans15/cwevansblindeconomistsswje2015-62366136

—— (2014), "Coins for Causes," Conscious Entrepreneurship Foundation White Paper (13 July). http://consciousentrepreneurship.org/coins-for-causes/

Fama, Eugene (1980) "Banking in the Theory of Finance," *Journal of Monetary Economics* 6(1), 39-57.

FinCEN (2013), "Application of FinCEN's Regulations to Persons Administering, Exchanging, or Using Virtual Currencies," *Guidance FIN-2013-G001*. http://www.fincen.gov/statutes_regs/guidance/html/FIN-2013-G001.html

Finley, Klint (2016-06-18), "A $50 Million Hack Just Showed That the DAO Was All Too Human," *Wired* [online]. https://www.wired.com/2016/06/50-million-hack-just-showed-dao-human/

Gabber, Eran, Markus Jakobsson, Yossi Matias, and Alain J. Mayer (1998), "Curbing Junk E-Mail via Secure Classification," *Financial Cryptography* (*Lecture Notes in Computer Science* 1465), 198–213.

Gans, Joshua S. and Hanna Halaburda (2013), "Some Economics of Private Digital Currency ," Bank of Canada Working Paper 2013-38. http://bankofcanada.ca/wp-content/uploads/2013/11/wp2013-38.pdf

Gerstein, Irving R. and Céline Hervieux-Payette (2015), "Digital Currency: You Can't Flip This Coin," Report of the Standing Senate Committee on Banking, Trade, and Commerce. Ottawa, Canada: Parliament of Canada. http://www.parl.gc.ca/Content/SEN/Committee/412/banc/rep/rep12jun15-e.htm

Graeber, David (2014), *Debt: The First 5,000 Years*, Brooklyn, NY: Melville House.

Graef, Jean (2010), "IT Neologisms: Necessary but Dangerous," *Customizing and Governing the SharePoint Search System*. Available from the *Montague Institute Review* (January) at: https://web.archive.org/web/20141012043254/http://www.montague.com:80/abstracts/neologisms.html

Greenfield, Robert L., and Leland B. Yeager (1983) "A Laissez-Faire Approach to Monetary Stability," *Journal of Money, Credit, and Banking* 15, 302–315.

Grinberg, Reuben (2011), "BitCoin: An Innovative Alternative Digital Currency," *Hastings Science & Technology Law Journal* 4, 159-208. http://hstlj.org/articles/bitcoin-an-innovative-alternative-digital-currency/
http://scienceandtechlaw.org/bitcoin-an-innovative-alternative-digital-currency/

Hall, Robert E. (1982a) "Explorations in the Gold Standard and Related Policies for Stabilizing the Dollar," *Inflation: Causes and Effects* (Robert E. Hall, Ed.), Chicago: University of Chicago Press, 111–122.

—— (1982b) "Monetary Trends in the United States and the United Kingdom: A Review from the Perspective of New Developments in Monetary Economics," *Journal of Economic Literature* 20, 1552–1556.

ICO Now (2018), "List of Bitcoin Forks," *ICO Now* https://iconow.net/list-of-bitcoin-forks/

IRS (2014), "IRS Virtual Currency Guidance : Virtual Currency Is Treated as Property for U.S. Federal Tax Purposes; General Rules for Property Transactions Apply," *Notice 2014-21*. http://www.irs.gov/pub/irs-drop/n-14-21.pdf

Jakobsson, Markus; Juels, Ari (1999), "Proofs of Work and Bread Pudding Protocols," *Communications and Multimedia Security* (*IFIP Conference Proceedings* 152), 258–272.

Kirk, Jeremy (2013), "Could the Bitcoin Network Be Used as an Ultrasecure Notary Service?" PC World 24th May. http://www.pcworld.com/article/2039705/could-the-bitcoin-network-be-used-as-an-ultrasecure-notary-service.html

Kocherlakota, Narayana R. (2002a), "Money: What's the Question and Why Should We Care about the Answer?" *American Economic Review* 92(2), 58-61.

—— (2002b) "The Two-Money Theorem," *International Economic Review* 43(2), 333-346.

—— (1998a), "Money Is Memory," *Journal of Economic Theory* 81(2), 232-251.

—— (1998b), "The Technological Role of Fiat Money," *Federal Reserve Bank of Minneapolis Quarterly Review* 22(3), 2–10.

Kocherlakota, Narayana R. and Neil Wallace. 1998. "Incomplete Record-Keeping and Optimal Payment Arrangements." *Journal of Economic Theory*, 81(2): 272-289.

Kristoufek, Ladislav (2013). "BitCoin Meets Google Trends and Wikipedia: Quantifying the Relationship between Phenomena of the Internet Era." *Scientific Reports* 3(3415), 1-7. http://www.nature.com/srep/2013/131204/srep03415/pdf/srep03415.pdf

Kroll, Joshua A., Ian C. Davey, and Edward W. Felten (2013), "The Economics of Bitcoin Mining, or Bitcoin in the Presence of Adversaries", *The Twelfth Workshop on the Economics of Information Security (WEIS 2013)*, Washington, DC, June 10-11 2013. https://www.cs.princeton.edu/~kroll/papers/weis13_bitcoin.pdf

Larimer, Stan (2013-09-14), "Bitcoin and the Three Laws of Robotics," *Let's Talk Bitcoin* http://letstalkbitcoin.com/bitcoin-and-the-three-laws-of-robotics/

Lessig, Lawrence (1999), "The Law of the Horse: What Cyberspace Might Teach," *Harvard Law Review* 113, 501-549. http://cyber.law.harvard.edu/works/lessig/finalhls.pdf

Levy, Steven (1994-12), "E-Money (That's What I Want)," *Wired* 2.12. http://archive.wired.com/wired/archive/2.12/emoney.html

Lo, Stephanie and J. Christina Wang (2014), "Bitcoin as Money?" Federal Reserve Bank of Boston *Current Policy Perspectives* 14-4, 1-28. http://www.bostonfed.org/economic/current-policy-perspectives/2014/cpp1404.htm

Luther, William J. and Josiah Olson (2015), "Bitcoin Is Memory," *Journal of Prices & Markets* 3(3), 22-33. http://pricesandmarkets.org/volume-3-issue-3-winter-2015/bitcoin-is-memory/

Lamport, Leslie, Robert Shostak, and Marshall Pease (1982), "The Byzantine Generals Problem," *ACM Transactions on Programming Languages and Systems* 4(3), 382-401.

Mark, Dino, Vlad Zamfir, and Emin Gün Sirer (2016-05-27), "A Call for a Temporary Moratorium on The DAO," *Hacking, Distributed.* http://hackingdistributed.com/2016/05/27/dao-call-for-moratorium/

Meikle, Scott (1994), "Aristotle on Money," *Phronesis* 39(1), 26-44.

Meiklejohn, Sarah, Marjori Pomarole, Grant Jordan, Kirill Levchenko, Damon McCoy, Geoffrey M. Voelker, and Stefan Savage (2013), "A Fistful of Bitcoins: Characterizing Payments Among Men with No Names," Internet Measurement Conference presentation, 23-25 October. http://cseweb.ucsd.edu/~smeiklejohn/files/imc13.pdf

Menger, Carl (1892), "On the Origin of Money," *Economic Journal* 2, 239–255. https://mises.org/library/origins-money-0

Metz, Cade (2016-06-16), "The Biggest Crowdfunding Project Ever—The DAO—Is Kind of a Mess," *Wired* [online]. https://www.wired.com/2016/06/biggest-crowdfunding-project-ever-dao-mess/

Miller, Mark S. and Marc Stiegler, 2003, "The Digital Path: Smart Contracts and the Third World," in Markets, Information and Communication: Austrian Perspectives on the Internet Economy (Jack Birner and Pierre Garrouste, Eds.) New York: Routledge: 63-88.

Moore, Geoffrey (1999 [1991]), *Crossing the Chasm*, New York: Harper Business.

Nakamoto, Satoshi (pseud.) (2008), "Bitcoin: A Peer-to-Peer Electronic Cash System," Working Paper, 1-9. https://www.bitcoin.com/wp-content/uploads/2017/01/bitcoin.pdf

Narayanan, Arvind, Joseph Bonneau, Edward Felten, Andrew Miller, and Steven Goldfeder (2016), *Bitcoin and Cryptocurrency Technologies*, Princeton, NJ: Princeton University. https://d28rh4a8wq0iu5.cloudfront.net/bitcointech/readings/princeton_bitcoin_book.pdf

NCSA (2013), "Enabling Discovery: NCSA Mosaic™," National Center for Supercomputing Applications (webpage). http://www.ncsa.illinois.edu/enabling/mosaic

Next! (1999), "Hoe DigiCash Alles Verknalde," *Next!* 27th April. http://web.archive.org/web/19990427142412/http://www.nextmagazine.nl/ecash.htm Translated as "How DigiCash Blew Everything," http://cryptome.org/jya/digicrash.htm

Owen, Oliver (2009), "Biafran Pound Notes," *Africa* 79(4), 570-594.

Pease, Marshall, Robert Shostak, and Leslie Lamport (1980), "Reaching Agreement in the Presence of Faults," *Journal of the Association for Computing Machinery* 27(2), 228-234.

Pluzhnyk, Angelina and Charles W. Evans (2014), "There's No Accounting for Bitcoin," Working Paper.

Popper, Nathaniel (2016-05-27), "Paper Points Up Flaws in Venture Fund Based on Virtual Money," *New York Time* [online].

Radford, R.A. (1945), "The Economic Organization of a P.O.W. Camp," *Economica* 12, 189-201. http://icm.clsbe.lisboa.ucp.pt/docentes/url/jcn/ie2/0POWCamp.pdf

Rahn, Richard W. (1999), *The End of Money and the Struggle for Financial Privacy*, Seattle: Discovery Institute.

Raskin, Max I. (2015), "Realm of the Coin: Bitcoin and Civil Procedure," *Fordham Journal of Corporate & Financial Law* 20(4), 968-1011. http://papers.ssrn.com/sol3/papers.cfm?abstract_id=2620309

Rivest, Ron, Adi Shamir, and Leonard Adleman (1978), "A Method for Obtaining Digital Signatures and Public-Key Cryptosystems," *Communications of the ACM* 21(2), 120–126.

Rogers, Everett M. (1962), *Diffusion of Innovations*, New York: Free Press.

Rooney, Ben (2011), "Women and Children First: Technology and Moral Panic," *Wall Street Journal* 11 July. http://blogs.wsj.com/tech-europe/2011/07/11/women-and-children-first-technology-and-moral-panic/

Rosenfeld, Meni (2012), "Overview of Colored Coins," Thesis. http://www.bibliothesis.com/thesis/2058001

Schneier, Bruce (1996), *Applied Cryptography* 2nd ed., Hoboken, New Jersey: John Wiley & Sons.

Seyfang, Gill and Ruth Pearson (2000), "Time for Change: International Experience in Community Currencies," *Development* 43(4), 56-60.

Simovitz, Aaron (2016), "Siting Intangibles," *New York University Journal of International Law and Politics* 46. http://papers.ssrn.com/sol3/papers.cfm?abstract_id=2456811

Standard & Poors (2018), "S&P 500," *S&P Dow Jones Indices*. https://us.spindices.com/indices/equity/sp-500

Szabo, Nick, 1997, "The Idea of Smart Contracts," Web Post. http://szabo.best.vwh.net/idea.html

Telegraaf (2013), "Drugsdealer Laat Zich Betalen met Lego-Dozen [Dutch: Drug Dealer Let Customers Pay Him with Lego Sets," *Telegraaf* (online, 2 August). http://www.telegraaf.nl/binnenland/21780673/__Drugsdealer_betaald_met_Lego__.html

Time (1995), "Cyberporn," *Time* 3 July, cover. http://img.timeinc.net/time/magazine/archive/covers/1995/1101950703_400.jpg

Tobin, James (1963), "Commercial Banks As Creators of 'Money', *Banking and Monetary Studies* (Dean Carson, Ed.), Homewood, Illinois: Irwin, 408-419.

United States District Court for the Northern District of California (2015), "United States of America v. Carl Force IV, et al." Criminal Complaint 25 March. http://www.justice.gov/usao-ndca/pr/former-federal-agents-charged-bitcoin-money-laundering-and-wire-fraud

Velde, François R. (2013), "Bitcoin: A Primer," Federal Reserve Bank of Chicago *Essays on Issues* 317, 1-4. http://www.chicagofed.org/webpages/publications/chicago_fed_letter/2013/december_317.cfm

Vickrey, William S. (1964), *Metastatics and Macroeconomics*, New York: Harcourt, Brace & World.

—— (1955), "Stability through Inflation," in *Post-Keynesian Economics* (Kenneth K. Kurihara, Ed.), London: George Allen and Unwin, 89-122.

Wheatley, Spencer, Didier Sornette, Tobias Huber, Max Reppen & Robert N. Gantner (2018-03-15), "Are Bitcoin Bubbles Predictable? Combining a Generalized Metcalfe's Law and the LPPLS Model," *ArXiv* 1803.05663. https://arxiv.org/abs/1803.05663

White, Horace (1902 [1895]), *Money and Banking* 2nd ed., Boston: Ginn & Co.

Wijk, Dennis van (2013), "What can be expected from the BitCoin?" Erasmus Rotterdam Universiteit Bachelor Thesis No. 345986. http://thesis.eur.nl/pub/14100

Young, Wallace (2015), "What Community Bankers Should Know about Virtual Currencies," *Community Banking Connections* 2nd Quarter. https://cbcfrs.org/articles/2015/q2/virtual-currencies

Zarate, Juan C. (2013a), "The Coming Financial Wars," *Parameters* 43(4), 87-97. https://www.scribd.com/document/276140478/The-Coming-Financial-Wars-by-Juan-C-Zarate

—— (2013b), *Treasury's War: Unleashing a New Era of Financial Warfare*, New York: PublicAffairs.

Appendix 1: Resources

General Information

Bitcoin.com: https://bitcoin.com
Bitcoin Wiki: https://en.bitcoinwiki.org/

Price Data

NYSE Bitcoin Price Index
    https://www.nyse.com/quote/index/NYXBT
Bitcoinity (time series plots and current order book)
    https://www.bitcoinity.org/markets/
Bitcoin Charts (current time series data dating back to 2012)
    https://bitcoincharts.com/charts/bitstampUSD
Bitcoin Investment Trust (OTC)
    http://finance.yahoo.com/q?s=GBTC
Coin Market Cap (current and historical prices for 100s of cryptocurrencies)
    https://CoinMarketCap.com

Bitcoin News Outlets

Bitcoin News: https;//news.bitcoin.com
Coin Desk: https://www.coindesk.com/
Coin Telegraph: https://cointelegraph.com

Online Bitcoin Discussions

Reddit: https://reddit.com/r/btc/
Bitcoin.com: https://forum.bitcoin.com/bitcoin-discussion

Bitcoin Wallet Software

https://www.bitcoin.com/choose-your-wallet/bitcoin-com-wallet

Bitcoin Exchanges

Coinbase: https://www.coinbase.com/
GDAX: https://www.gdax.com/
Gemini: https://gemini.com/
Kraken: https://kraken.com/

Appendix 2: Remittances

One of the earliest uses for Bitcoin that did not involve its use as a medium of exchange for goods and services was for international remittances. The most basic form involves a remittor's acquiring bitcoins and transferring them to a remittee, who then sells them for local currency. A somewhat more complex system involves a third party who receives the bitcoins on behalf of the remittee, sells them, and then delivers the proceeds to the remittee.

An even more sophisticated system involves one like BTC Ghana (https://btcghana.com/). Here, the remittor acquires bitcoins and transfers them to BTC Ghana. BTC Ghana sells the bitcoins for Ghanaian cedi and purchases mobile telephone credit on the Tigo Cash, Airtel Money, or MTN Mobile Money network, and then transfers the mobile phone credit to the remittee, who then uses the mobile telephone credits as money at shops in Ghana.

An Internet search for "Bitcoin" and "currency controls" yields numerous stories about how Bitcoin has been used to get money out of countries like Argentina, Greece, and Venezuela, as well as remit money to friends and relatives in those countries. As mainstream cryptocurrency adoption increases worldwide, this kind of application is expected to grow.

Appendix 3: Bitcoin in Islamic Banking and Finance

The discussion below was published as "Bitcoin in Islamic Banking and Finance," in the *Journal of Islamic Banking and Finance* in 2015. Although it might be unusual for a forensic economist in the USA to be called upon to testify as an expert witness on cryptocurrency in a civil or criminal case involving Islamic banking and finance, it nonetheless illustrates one of the potentially unexpected ways, in which cryptocurrency enables financial arrangements that are not otherwise cost effective, or might even be impossible.

The term *Blockchain Management System* never caught on. It is essentially synonymous with *cryptocurrency*.

A3.A. Abstract

This Appendix analyzes the compliance of distributed, autonomous blockchain management systems (BMS) like Bitcoin—also referred to as 'virtual currencies'—with the requirements of Islamic Banking and Finance. While intended as a narrow financial and economic analysis, and not as an in-depth analysis of the subtleties and nuances of *Shari'a* as they relate to banking and finance, it shows that a BMS can conform with the prohibition of *riba* (usury) and incorporate the principles of *maslaha* (social benefits of positive externalities) and mutual risk-sharing (as opposed to risk-*shifting*). It concludes that Bitcoin or a similar system might be a more appropriate medium of exchange in Islamic Banking and Finance than *riba*-backed central bank fiat currency, especially among the unbanked and in small-scale cross-border trade.

Keywords: Islamic Banking, Islamic Finance, Bitcoin, virtual currency

A3.B. Introduction

Ariff (2014) notes that, although the literature on modern Islamic Banking and Finance (IBF) dates back more than a half-century, the practice of IBF is still in its infancy, and some common practices of Islamic banks are "questionably asymmetrical" and "in consonance with the letter rather than the spirit" of Islamic traditions (p.741). He suggests that this is an understandable consequence of the fact that this initial phase of IBF focuses on offering *Shari'a-compliant* alternatives, and that the next phase should focus on *Shari'a-based* services that might bear little or no resemblance to conventional banking and financial services.

Even with these initial shortcomings, small Islamic banks tend to enjoy lower credit and insolvency risk than their conventional counterparts, and the loan quality of both small and large Islamic banks appears to be less responsive to changes in domestic interest rates than the loan quality of conventional banks operating in the same jurisdictions. (Abedifar, Molyneux & Tarazi 2013)

This paper argues that a distributed, autonomous blockchain management systems (BMS) like Bitcoin—also referred to as a 'virtual currency'—might find a place within this growing field that is increasingly popular among not only the 20-25% of the world's population who are Muslims, but among many non-Muslims, as well (Abidefar, et al. 2013; El-Gamal 2006; Ghannadian & Goswami 2004; Hasan & Dridi 2010; Imam & Kpodar 2014; M.M. Khan & Bhatti 2008). This paper begins with a brief review of modern IBF literature and background on BMS. it concludes with a discussion of the potential relationship between the two.

A3.C. Islamic Banking and Finance

Although the literature on modern IBF dates back more than a half-century, IBF is still a work-in-progress. Disagreement continues among proponents and outside observers concerning the distinction between what is permitted or required (*halal*) and what is forbidden (*haram*) under different interpretations of Islamic Law (*Shari'a*). Adding to the confusion is the desire of many Islamic banks' executives and customers to integrate those banks into the global banking and financial system, which is governed by international treaties and national statutes and regulations that often are at odds with or even in direct violation of *Shari'a*. (Ariff 2014, 1988)

Nonetheless, it is universally recognized that the overriding feature of IBF is the prohibition of usury (*riba*). Granted, some heterodox Islamic scholars argue for a distinction between exploitative usury and benign time preference in the form of interest on money loans, or for a distinction between consumer debt and commercial debt, but the consensus among IBF scholars is that all transactions should involve the transfer of real goods and services (Ariff 1988), echoing Aristotle's position of seeing money as a facilitator of the trade of real goods and services—not a good in its own right—and holding 'unnatural' money-for-money loans in disdain, especially the most hated sort: the breeding of money from money (Meikle 1994).

Readers from the West/Global North who find this position odd are reminded that restrictions on economic activity are common worldwide in the forms of licensing, regulation, patents and copyrights, government franchise, civil forfeiture, and prohibitions of usury (i.e., 'excessive' interest rates), unconscionable contracts, insider trading, front-running, price gouging, Ponzi and pyramid schemes, money laundering, *suspected* money laundering, etc. In this way, the prohibition of *riba* should be no more exceptional than, e.g., the prohibition of trade based on knowledge of fact (insider trading, front-running, price gouging), of providing services that are otherwise legal (government franchise, licensing), or of sharing information (patent, copyright). In this spirit, let us look at IBF as it currently is practiced.

M.N. Siddiqi's (1988 [1969]) seminal *Banking without Interest* states that a bank should be organized as a *Shirkat-e-*

Enan, eschewing *riba* and earning income from profit-sharing (*mudarabah*) and profit- and loss-sharing (*musharakah*, also known as *shirkah*). In conventional Western/Northern terms, this is essentially a call for the organization of banks as general partnerships that consist of at least two partners, both or all of whom bear unlimited liability, sharing profits *and* losses in proportion to the relative sizes of their investments. While all partners should have equal rights to participate in the daily operation of the bank, they need not be required to do so. Individuals can elect to be passive—though not *limited*—partners, who delegate operations to a subset of the partners or to administrators who are not partners, and they can have business interests outside the bank that are unrelated to their participation in the bank.

The primary concern is to avoid principal/agent asymmetries that exist in conventional limited-liability firms, especially in jurisdictions with efficient capital markets that make it trivially easy for shareholders to sell their shares, rather than expend the effort of monitoring board members and officers (executives). Given a choice between monitoring executives on the one hand, calling them to task when they misbehave, and struggling to change their behavior, and simply selling one's shares on the other hand when executives misbehave, shareholders generally opt for the latter, where capital markets are efficient. A very large literature exists on how this principal/agent problem results in executives of publicly traded firms being free to pursue their own interests, which can run contrary to those of the shareholders. (Ariff 2014, 1988; Siddiqi 1988 [1969])

While many modern Islamic banks are organized as conventional limited-liability corporations with passive shareholders, this is generally seen as a shortcoming to be remedied. Something more closely resembling a credit union or a cooperative is preferable, in which members are owners in proportion to the relative sizes of their investments in the venture, and all owners have an incentive to monitor executives actively. This remains an area for future research and exploration. (Ariff 2014, 1988)

Choudhury & Hussain (2005) emphasize the centrality of the promotion of the precepts of *Shari'a* in the practice of IBF, including economic development, human capital, social justice, etc. Ariff (1988) sees this as a secondary concern that derives its rationale from the eschewing of *riba*. Even so, citing Chapra (1982), he notes that the scope of activities that Islamic banks undertake need not mirror those of conventional banks, and can include a different mix of services with a stronger emphasis on social welfare than profit maximization. Similarly, it is not uncommon in the West/Global North for fiscal policy makers to issue mandates to bank executives to extend credit on the basis of economic development, human capital, and social justice, including guaranteed student loans, subsidized mortgages, sub-market and guaranteed loans to businesses in economic development zones, etc.

While one might see economic development, human capital, and social justice as distractions from profit maximization, Taleb (2012, 2007, 2005) emphasizes that the single-minded focus on profit maximization can lead to excessive financial leverage (gearing) and over-optimization, which can cause losses over short intervals that are greater in magnitude than total profits over the preceding decades. A. Ahmed (2010), Chapra (2010), Hasan & Dridi (2010), M.F. Khan (2010), Matthews & Tlemsani (2010), Seidu (2010), M.N. Siddiqi (2010), Smolo & Mirakhor (2010) echo this point in their analyses of the relatively mild effects of the Crash of 2008 on Islamic banks, as contrasted with its nearly devasating impact on American and European banks.

Readers interested in more detailed introductions to IBF are referred to Aghnides (1916), Z. Ahmed, Iqbal & Khan (1983), Ariff (1982), Obaidullah (2005), A. Siddiqi (2006), and M.N. Siddiqi (1988 [1969]).

With regard to the broader, general topic of the influence of religion on economics and finance, Weber (2001 [1905]) describes in detail how the understanding of modern economics in the West/Global North and the very form of capitalism itself are infused with Protestantism and Utilitarianism. This influence of religion on culture and the economic systems that operate within cultures led Weber to publish *The Religion of China: Confucianism and Taoism* (1968 [1915]), *The Religion of India: The Sociology of Hinduism and Buddhism* (2012 [1916]), and *Ancient Judaism* (1967 [1917-1919]); and he was continuing his series with works on early Christianity and Islam, but was cut short by his unexpected death in 1920 from pneumonia as a complication of Spanish flu (Bendix 1977 [1960]).

Far from originating in the physical laws of nature—like the need to eat and sleep, the principles of stable public works construction, or the practice of animal husbandry and agriculture—economic organization everywhere and among all peoples originates in custom and tradition, which are influenced by religion (Lavoie & Chamlee-Wright 2000). In this regard, and setting aside theological differences, IBF in practice is different from conventional banking in form, but not necessarily in kind, at this stage of its development. This could change, however, if Islamic bank executives abandoned inflationary *riba*-backed central bank money (fiat) in favor of something more harmonious with IBF.

A3.D. Blockchain Management Systems

Conceptually, a blockchain management system (BMS)—the most successful being Bitcoin—is a self-contained system for transferring numerical values from one account to another, such that no value is lost in transit between accounts, and double-spending is impossible. In this way, a BMS can be seen as an accounting system.

BMSs like Bitcoin exist as myriad copies of a piece of software that run on users' computers, communicate with each other over the Internet, and have copies—that are updated approximately every ten minutes—of the history of every transaction that has been completed within the system since its inception. If anyone's transaction history differs from others' it is considered to be incorrect and it is replaced with a copy of the correct record. To subvert the system, one would need to control more than half of the entire network and to corrupt the record in precisely the same way across that majority. (Nakamoto 2008)

When discussing Bitcoin, the capitalized term (*Bitcoin*) refers to the software and the network of users, and the lowercase (*bitcoin*) refers to the *unit of account* in the system. Here, XBT—following the standard for currency abbreviations defined by ISO 4217 (International Organization for Standardization 2008)—is used interchangeably with *bitcoin*.

Pluzhnyk & Evans (2014) approach the Bitcoin system as a provider of transaction confirmation services and treat XBT as liabilities of that service provider that are used as a *medium of exchange* and sometimes as a *store of value*. Given that the system is self-contained, all of the conventional balance sheet accounts—on the asset side: cash, accounts receivable, inventory, fixed assets; on the liabilities side: accounts payable, notes payable, long-term debt, paid-in capital—are equal to zero. Nonetheless, markets for XBT are active 24/7 with no breaks for holidays, and the market cap for Bitcoin can be calculated from the most recent price and the total amount of XBT in existence. This value currently is measured in the billions of USD.

Considering that these XBT liabilities have market value, it holds that the Bitcoin system must hold some asset of equivalent value. Pluzhnyk & Evans (2014) invoke *goodwill* to describe this intangible asset, and argue that it derives from users' expectations of Bitcoin's usefulness in future transactions. Evans (2015) notes that the current market value (*P*) of XBT is driven by users' continually updated expectations concerning three factors: the value of future services and cost savings (*F*) enabled by Bitcoin; the market consensus of the appropriate discount rate (*r*), which includes ever-changing premiums for myriad sources of perceived risk; and the time (*T*) until *F* is realized, such that:

$$P = \frac{F}{(1+r)^T}, \text{ where:}$$

- **F** includes the value of users' ability 1) to transact directly with the billions of individuals worldwide who do not have easy access to efficient banking and money transfer services, but own mobile phones, 2) to buy from online retailers at Chinese or US prices, plus shipping, rather than at the substantially higher local prices that prevail in many parts of the world, 3) to remit any amount to anyone anywhere in the world who has Internet access, 4) to hold value privately, particularly in areas where holders of large bank account balances or recipients of large transfers via Western Union are in danger of being kidnapped for ransom. As realization among market participants of these advantages increases, *F* increases, putting upward pressure on *P*.

- **T** is the time for useful innovations to come to market. As expectations of the delay until they become available decrease, *T* decreases, putting upward pressure on *P*.

- *r* includes premiums for the risks of technical difficulties and security issues related to holding XBT, regulatory uncertainty, price volatility, etc. As uncertainty decreases, *r* decreases, putting upward pressure on *P*.

Abidefar, et al. (2013), Ariff (2014, 1988), Choudhury & Hussain (2005), El-Gamal (2006), Obaidullah (2005), and A. Siddiqi (2006), *inter alia*, remind readers that time value of money calculations, *per se*, are not *haram*, so long as the differential between *P* and *F* is associated with the purchase of real assets, and both parties to the transaction share the risk. For example, it is permissible to accept a lower price for the pre-sale of an asset delivered later and paid for now, or to require a higher total amount paid for an installment purchase delivered now and paid for over time, so long as both parties to the transaction share the risk.

It is beyond the scope of this paper to explore Bitcoin's compliance with *Shari'a* as a *store of value*. If it is seen as an undifferentiated pre-payment for goods and services in the indefinite future that have not yet been identified, this might lead to a different conclusion from the assumption that XBT is a new kind of digital asset that can be used as money. However, if XBT is used as a *medium of exchange* for transactions completed within very short time spans, then Bitcoin can be seen as a communication protocol, not entirely unlike an order for a bank transfer, a message sent between two *hawala* agents, or a cheque that is deposited immediately upon receipt. The exploration of these and related questions is a potentially productive wellspring for future research.

A3.D.i. The Regulators and the Elephant

Regarding the points made immediately above, we are reminded of the story of the Blind Men and the Elephant, in which each blind man generalizes that the elephant in its totality is represented by the part that he feels. To the one blind man touching a leg, and elephant is like a tree; to another near its trunk, it is like a snake; to yet another by its tail, it is like a rope; by its ear, a fan; by its ribs, a wall; etc. In a case involving forked cryptocurrencies, it is not blind men, but regulators confronted with something unfamiliar and innovative.

When we encounter new technology, the language that we use to describe it often fails us. We must use existing words in strange ways to describe new concepts associated with the innovation, in the same way that our ancestors had to, when confronted with the 'horseless carriage' (automobile), 'moving picture' (cinematic film), or 'information superhighway' (Internet) (Graef 2010). Today, it is 'virtual currency' (Bitcoin).

This leads to our using terms like 'coins' and 'mining' to describe how new XBT units come into circulation, although there are no *coins* in Bitcoin. Even if there were, coins are not mined; ore is mined, ingots are smelted from ore, and coins are minted from ingots. The terms 'coin' and 'virtual currency' are metphors. (Evans 2014a)

In the USA, to FinCEN (2013) officials, Bitcoin looks like a money transmission system. To Internal Revenue Service (2014) and Commodity Futures Trading Commission (2014) officials, XBT looks like a commodity. To some theorists, including Pluzhnyk & Evans (2014), XBT can be seen as non-voting liabilities of a 'distributed autonomous company' (Larimer 2013). To the degree to which the Bitcoin system provides the services that Fama (1980, p.39) describes as "the the main function of banks in the transactions industry," to wit: "the maintenance of a system of accounts in which transfers of wealth are carried out with bookkeeping entries," it is a bank, and XBT are liabilities of that bank. Countless Bitcoin supporters insist that XBT is a community currency (Blanc & Fare

2013; Evans 2014b; Owen 2009; Seyfang & Pearson 2000). Whereas, central bank researchers conclude that XBT looks *unlike* a currency (European Central Bank 2012; Lo & Wang 2014; Velde 2013).

Such broad disagreement is understandable, considering that regulators from specific ministries or regulatory agencies oversee specific subsets of regulated activities—i.e., banking, consumer credit, mortgage lending, equipment financing, venture funding, etc.—and financial innovations often blur the lines dividing one kind of activity from others. This can lead to disagreement over which ministry or regulatory agency has jurisdiction over the innovation, and it is only natural that regulators might prefer to view Bitcoin in terms of some preexisting category (Brito & Castillo 2013). The unlikely alternatives to this would be at the one extreme for regulators to petition legislators to draft, debate, and enact new statutes that create new ministries or regulatory agencies to oversee the use of this financial innovation by the public, or at the other extreme—in this post-9/11, post-USA PATRIOT Act, post-Crash of 2008 world—to disavow responsibility and let market participants sort it all out in a state of *laissez faire* (Zarate 2013). The former is costly and tedious, the latter runs counter to the regulatory instinct, and both dilute the power and influence of the authorities and agents employed by the ministry or regulatory agency shrugging off responsibility. As Brito & Castillo (2013) note, the outcome that we have seen among regulators in OECD member states is a tendency to put Bitcoin-related activities into preexisting regulatory 'buckets'.

Within the context of IBF, *Shari'a* authorities face the same dilemma as Western/Northern regulators. Although narrowly defined aspects of Bitcoin fit into neat categories, the totality of Bitcoin—or any BMS—transcends conventional categories, and it falls on authorities to proclaim which categories take precedence over others.

To see why this could be problematic, consider that Bitcoin-the-software is much more than the XBT bitcoins-the-units. XBT simultaneously can be seen as virtual currency liabilities of a kind of bank or central bank, as non-voting capital shares in the system as a whole, *and* a new class of commodity. XBT can be used as a *medium of exchange* and as a *store of value*.

Complicating things further, a programmer in Argentina has deveoped a way to embed messages into the Bitcoin blockchain, as a kind of time stamp that serves a very similar function as 'notarization' (Kirk 2013). Rosenfeld (2012) has created a way to tag specific units of XBT with extra strings of data (colored coins) that give them a unique 'color', that can mark them as proxies for shares of equity, bonds, future or forward contracts, options, leases, annuities, or any other financial assets. And, developers continue to develop cleverer and more surprising uses for this platform that can be reached from anywhere in the world with Internet access, and can be used for all manner of transactions and services that are *de jure* regulated and now *de facto* unregulatable.

This situation creates a conundrum for regulators. If officials from several ministries or regulatory agencies within a given jurisdiction—banking, stock exchanges, derivatives markets, money transmission, taxes, consumer protection, etc.— see this metaphorical 'elephant' as being within their specific purviews, then to set it free in an state of *laissez faire* would call into question the rationale of the regulatory state. If, instead, regulators within specific ministries or regulatory agencies tried to apply existing regulations to this innovative technology that is both a superset of existing categories and oblivious to jurisdictional boundaries, then they would risk chasing it underground, where it would be beyond their reach, and they would create incentives for Bitcoin developers and entrepreneurs to emigrate to more liberal jurisdictions, taking their startup firms—perhaps the next generation's 'Apple', 'Facebook', 'Google', etc.—with them, as appears to be happening in the USA already (Torpey 2014).

The response among regulators in OECD member states has been to make peace with virtual currency. In some jurisdictions, particularly within the EU, this has been enthusiastic. In the USA it has been begrudging, with officials at the CFTC (2014), FinCEN (2013), and the IRS (2014) issuing positive signals, while officilas at the Department of Justice issue negative signals and officials at the Office of the Comptroller of the Currency issue *no* signals. This creates enough fear, uncertainty, and doubt among Bitcoin entrepreneurs to keep them from completely outrunning regulators in the short run.

Although *Shari'a* and Anglo-Saxon Common Law are orthogonal to each other, policy makers operating within both legal traditions face the same technical reality. The following observations are offered in this spirit.

A3.D.ii. What *Is* Money?

Money, as an institution is as old as civilization. The literature on money dates at least as far back as Aristotle (Meikle 1994; Menger 1892; White 1902), and many variations on the story of the evolution from tribal communism to barter to money have been told over the ages. A commonly accepted definition of money is along the lines of: *a commodity that individuals accept voluntarily in exchange for all other goods*. Historically, the commodity has tended to be a precious metal, usually gold (*dinar*) or silver (*dirham*). Granted, other things have served this purpose in isolated communities, but those are anthropological curiosities more than relevant examples here (Del Mar 1895). However, even in modern times, the drive to reckon in terms of some form of money is so strong that even cigarettes can serve as money, when conventional forms of money are not available (Radford 1945).

Historically, money, *per se*, served four functions: *medium of exchange, unit of account, store of value*, and *measure of value*. However, Cronin (2012) and Evans (2014a) demonstrate that these functions are generally separated in practice today, which leads to the conclusion that we might be witnessing the "end of money" (Rahn 1999). Nonetheless, the ideal commodity to serve as money has been understood throughout history to be fungible, portable, scarce, divisible, and durable. Gold and silver qualify in this regard, as proven by history, and—as discussed below—so does Bitcoin, although fiat does not.

Which of these qualities takes priority over the others is a matter of perspective. Within the context of IBF, a *Hadith* of Sahih Muslim (Book 10, Ch. 36) that addresses the use of money and the avoidance of paying or receiving *riba*, suggests that fungibility is as important as scarcity and durability. It admonishes individuals to exchange of like quantity for like quantity—gold for gold, barley for barley, dates for dates, etc. —and never more for less.

This presents an intriguing conundrum for holders of fiat, which tends to lose value over time. On the one hand, when one borrows, e.g., US dollars (USD) in a *riba*-free environment, one is expected to repay the same amount of USD, no more and no less. On the other hand, when one repays the same quantity of USD after some period of time, the fiat's purchasing power generally has fallen; if the period of the loan is long, then the value might have fallen significantly. The same *number* of USD in the future is worth less *in terms of the goods and services that it can buy* than that number of USD was worth in the past.

The most straightforward way to avoid this conundrum is to prohibit loans of money for more than a very short period. However, it is instructive to note that this specific issue developed only relatively recently with the institution of pure fiat worldwide beginning in the 1970s. Historically, this concern and the concomitant contradictions that it causes were moot.

Before the creation of modern, floating central bank money in the early 1970s, economists distinguished between *real money* and *promises to pay*, noting that the difference between them is analogous to the difference between a meal and a meal ticket, or a coat and a coat check (White 1902); today, one might compare a valet ticket with a car in this context. Even so, *riba*-backed Federal Reserve Notes (FRN) were declared in 1913 to be legal tender in the USA, after which they traded at parity with gold certificates, and their value was pegged indirectly to the value of gold. In this way, both gold certificates and FRNs served as promises to pay either real money or an equivalent value of other goods and services.

In 1933 it became a federal crime for US persons to own USD—then defined as $1/35^{th}$ of one troy ounce of gold—but the value of gold certificates and FRNs continued to track the value of gold. In 1973 the USD ceased to exist as a fixed quantity of gold and instead has become a *promise to pay* $1.00, backed by a loan from the Federal Reserve with a face value greater than $1.00, making $1.00 a *promise to pay **more than** $1.00*, itself being a *promise to pay*. Given that each unit of fiat—whether USD, euros, or any other—is lent into existence and backed by an obligation to repay *more* than one unit of that same currency in the future, the quantity of fiat must increase indefinitely, in order to avoid defaulting on the loans that backs each unit in circulation. It is a literal impossibility for all of the loans outstanding that back the fiat in circulation at any time to be repaid using fiat. This is similar, at the national level, to an individual who is caught in a spiral of debt that he or she cannot repay, and must borrow increasingly more, in order to make payments on existing debt.

Fiat is neither a long-term *store of value* nor a long-term *measure of value*. For example, if one had put gold certificates or FRNs with a face value of $35.00 into an envelope in the early 1970s, when the ban on gold ownership in the USA was lifted, on the expectation that the $35.00 could buy one troy ounce of gold forty years later (i.e., today at the time of this writing), one would have been disappointed. In terms of gold, more than 95% of the value that those $35.00 represented has vanished (*store of value*), and $35.00 *worth of goods* then was substantially more than $35.00 *worth of goods* today (*measure of value*). Granted, individuals continue to use fiat as a *medium of exchange*, but demand for cash and cheques is falling in the West/Global North, as individuals rely ever more on debit cards, credit cards, and online banking (Cronin 2012). It is not inconceivable that fiat eventually could serve merely as a *unit of account*, while other assets and services perform the functions of *medium of exchange*, *store of value*, and *measure of value*. (Black 1970; Fama 1980; Greenfield & Yeager 1983; Hall 1982a, 1982b; Rahn 1999)

A3.D.iii. BMS *versus* Central Bank

If we view a BMS like Bitcoin as a kind of central bank that issues a private currency, we can contrast XBT and fiat within the context of IBF. To appreciate how this can be, it helps to review briefly the process by which XBT are released into circulation.

In Bitcoin parlance, *mining* is an automated and decentralized form of transaction confirmation that maintains a stable and decelerating rate of the release of XBT into circulation. The ultimate supply of XBT is 21 million, hard-coded into the Bitcoin software, and expected to be reached sometime around the year 2040. The terms *miner* and *mining* are somewhat misleading, as they are intended to convey the technical rationale behind Satoshi Nakamoto's (2008) choice of the method by which the XBT are released into circulation, rather than the economic rational.

In the early, experimental days of Bitcoin, all user software mined, however this function has been turned off in later versions of the consumer and merchant software. Now, a minority of users—who have elected to install and operate application-specific integrated circuit (ASIC) chips that are able to perform only the function of confirming Bitcoin transactions— serve the role of miners. Anyone can buy ASICs and set up a mining operation, but because the total computing power of the Bitcoin network is several times as great as that of the top 500 supercomputers in the world combined (Matonis 2014), one would need to invest very heavily in equipment and Internet access in order to compete with existing miners.

Approximately every ten minutes the most recent Bitcoin transactions are bundled together automatically by the software network into a *block*, and all miners commence a race to see who can be the first to confirm the validity of the transactions in the block and then to solve a fiendishly difficult mathematical problem that has a random outcome, the correctness of which is easy to verify. The purpose of this step is to control the rate at which blocks are generated. When the confirmation rate becomes substantially shorter than ten minutes as miners add more computing power to the network, the software automatically increases the difficulty of the mathematical puzzles, so that the cycle stays as close as possible to one block every ten minutes.

The winner of a given block is decided largely by chance, but over the long run, the likelihood that a miner will be the first to confirm a particular block is in proportion to that miner's investment in ASICs. In this way, *reward and cost are symmetrical*. The miner that confirms a particular block first is awarded a fixed amount of XBT released by the software. The block is appended to the collection of previous blocks into the *blockchain*, all users who maintain their own copies of the blockchain update their copies, and the race begins again.

Because miners must pay for electricity, Internet access, and equipment maintenance, they typically sell some of their newly awarded XBT for fiat and keep the remainder as their profit or invest it into new ASICs to expand their operations. Thus, *newly released XBT are payment for miners' efforts to maintain the integrity of the network*. Miners expend real resources in pursuit of XBT, and they are rewarded in proportion to their investments. If the market price of XBT rises, they gain; if it falls, they lose.

Seen the other way around, from the perspective of the Bitcoin system, *per se*, if the miners collectively win, the system wins, because the incentive for others to become miners increases, which strengthens the system; if the miners collectively lose, the system loses, because the incentive to quit mining increases, which weakens the system. In this way, miners individually and the system, *per se*, share gains and losses.

Users who are not miners can acquire XBT either by buying them with fiat from someone who offers them for sale, or to provide goods or services in exchange for XBT. This is precisely the way that one acquires, e.g., euros if one is outside the Eurozone: buy them or earn them. If one acquires XBT, in order to effect a transaction in the short term, then one bears very little risk; one acquires XBT, uses it as a *medium of exchange*, and receives the good or service purchased. If, instead, one holds XBT for a long time as a *store of value*, then one bears the risk of the market price's falling, which is true of gold, silver, land, etc. as a *store of value*.

Once released into circulation, XBT's value derives from users' expectation of its ability to enable transactions more efficiently than conventional alternatives, or that have been hitherto prohibitively expensive or even impossible. Before the release of the first Bitcoin software, if one wanted to transact across national borders with individuals in the Emerging Middle Income Regions of the world—perhaps to buy a container of green coffee beans directly from a small-scale grower in Central America or Sub-Saharan Africa—one had to use the services of a bank or a money transmitter like Western Union. However, the World Bank (2012) estimates that more than 2.5 billion adults in the world do not have bank accounts, and money transmitter fees are often prohibitive. With Bitcoin, anyone who has Internet access via either a computer or a smartphone can transact directly with trading partners worldwide, while communicating in real time as the system confirms the transaction.

A3.D.iv. XBT *versus* Fiat

Contrast the creation of XBT with the creation of fiat:
- XBT is free of *riba*, whereas fiat is lent into existence in exchange for *riba*.
- The face value of the goodwill backing XBT is, by definition, *exactly equal to* the value of the XBT in circulation, whereas the face value of loans backing fiat is, by definition, *greater than* the value of the fiat in circulation, due to present value discounting; the excess value represents the issuing bank's equity, thereby granting it a perpetual advantage over fiat holders.
- New units of XBT come into circulation in exchange for the expending of real resources to maintain the integrity of the Bitcoin system, whereas new units of fiat come into circulation at the will of individuals who borrow them into existence, whenever they deem it desirable to do so.
- The quantity of XBT in circulation increases at a predictable and decelerating rate until it reaches its ultimate cap of 21 million units in circulation—each divisible to the $1/100$ million[th]—whereas the ultimate total quantity of fiat is unlimited and can hyperinflate.
- XBT is backed by the expectation that it will enable transactions for real goods and services that hitherto have been prohibitively expensive or even impossible, whereas fiat is backed by debt *denominated in itself* with a face value greater than the total amount of fiat in circulation.

A3.E. Modest Proposals

Ariff (2014) notes that IBF, as it is practiced today, is still in its infancy, and that the focus still is on mimicking the services of conventional banks, but in a *Shari'a-compliant* manner. However, Ariff (1988), citing Chapra (1982), notes that Islamic banks need not mirror conventional banks, and in the second phase of the evolution of IBF, they might include a different mix of *Shari'a-based* services.

In this spirit, the following speculations and hypotheses are offered for discussion:

- The executives of a group of Islamic Banks—either an affiliated group or a loose confederation—could organize a virtual currency exchange under the principal of *musharakah*, in order enable those banks' customers to buy and sell XBT efficiently, in order to transfer value amongst themselves and to bypass the inefficiencies of the *status quo* banking system. If this exchange maintained a very narrow bid/ask spread and charged no other fees, and restricted access to customers of the member banks, this could create an incentive for Muslims and non-Muslims alike to bank with member banks.

- For non-Muslims who appreciate the greater stability of a banking system that is free of incentives to over-optimize and over-leverage, the set of *halal* financial services can be approximated with a collection of conventional non-bank financial services. Such an 'Islamic' 'bank' might weather future crises in the conventional fiat banking system relatively unscathed, in similar fashion to how small Islamic banks survived the Crash of 2008 relatively unscathed. These services include trust and escrow, real estate and equipment leasing, venture funding, business brokerage, and potentially a credit union. In many jurisdictions, the startup costs and reserve requirements for each of these services combined can be a fraction, perhaps by even one or two orders of magnitude, of the cost of acquiring a banking license.

• In an ideal world, Bitcoin would not be necessary, and Islamic banks could transact in gold. However, doing so could isolate a gold bank from the global banking system within a financial sandbox, and 'virtual gold' has been created by several groups of entrepreneurs since the 1990s with dubious results. Bitcoin, on the other hand, does not require vaults, guards, custodians, and other very expensive points of failure based on trusted individuals whom one does not know. Also, if one wants to transact using gold across national borders, transportation and security can be prohibitively expensive. If one resorts to *hawala*-style networks with trusted agents at each end, then one winds up where we are today. At the other extreme, one can use the *status quo* banking system and fiat, which are founded on principals that violate *Shari'a*. While Bitcoin might not be perfect, one could argue that it is less bad than current alternatives.

A3.F. Concluding Discussion

This paper analyzes the relationship between a distributed, autonomous blockchain management systems (BMS) like Bitcoin —also referred to as a 'virtual currency'—and Islamic Banking and Finance (IBF). It shows that a BMS can conform with the prohibition of *riba* (usury)—as Bitcoin does—and incorporate the principles of *maslaha* (social benefits of positive externalities) and mutual risk-sharing (as opposed to risk-*shifting*).

With regard to *maslaha*, the world's unbanked number in the billions and represent the majority of the world's adults. Those among them who have Internet access—especially those who own smartphones—can use Bitcoin to transact as equals in the increasingly integrated global marketplace, bypassing inefficient banks, rapacious money transmitters, and multiple layers of wholesalers, cooperatives, and other intermediaries who extract markups that could accrue to the original producers. Also, if a Bitcoin user takes positive steps to promote its benefits to the world at large, then the value of his or her XBT should increase, thereby benefiting all Bitcoin users collectively.

With regard to risk-sharing and cost-sharing, at its foundation Bitcoin operates according to *musharakah*, in which 'miners' operate as general partners in loose confederation, who share the costs and benefits of maintaining the system. The greater in proportion to the total computing power in the system a particular miner's investment is in equipment used to confirm transactions among users, the greater is that miner's proportion of the XBT released by the system. The more miners there are in the system, both in terms of population and computing power, the more secure the system is, and the greater the value of the XBT that they receive for their services.

When the origin of a unit of XBT is compared with the origin of a unit of fiat, the contrast is stark. Fiat is born of *riba* and, by definition, not enough exists to repay the loans backing it; it necessarily must inflate without end. With Bitcoin, the quantity of XBT in circulation increases at a predictable and decelerating rate, and the ultimate cap on the quantity is hard-coded into the software at 21 million units. Rather than being lent into existence at the whim of an individual, a new unit of XBT is released in exchange for the provision of services to maintain the security and stability of the Bitcoin network.

Supporters of both IBF and XBT prefer hard money over fiat. Potentially useful things can be achieved, if members of these two very different communities become acquainted.

A3.G. References

Abedifar, Pejman, Philip Molyneux & Amine Tarazi (2013), "Risk in Islamic Banking," *Review of Finance* 17, 2035–2096.

Aghnides, Nicholas P. (1916), *Mohammedan Theories of Finance*, New York: Columbia University.

Ahmed, Adel (2010), "Global Financial Crisis: An Islamic Finance Perspective," *International Journal of Islamic and Middle Eastern Finance and Management* 3(4), 306-320.

Ahmed, Ziauddin , Munawar Iqbal & Mohammad Fahim Khan, eds. (1983), *Money and Banking in Islam* , Jeddah: Islamic Economic Research Center (King Abdulaziz University).

Ariff, Mohamed (2014), "Whither Islamic Banking?" *The World Economy* 37(6), 733-746.

—— (1988), "Islamic Banking," *Asian-Pacific Economic Literature* 2(2), 48-64.

——, ed. (1982), *Monetary and Fiscal Economics of Islam*, Jeddah: Islamic Economic Research Center (King Abdulaziz University).

Bendix, Reinhard (1977 [1960]), "Max Weber's Sociology of Religion," in *Max Weber: An Intellectual Portrait*, Berkeley: University of California, 257-284.

Blanc, Jérôme and Marie Fare (2013), "Understanding the Role of Governments and Administrations in the Implementation of Community and Complementary Currencies," *Annals of Public and Cooperative Economics*, 63–81.

Black, Fischer (1970) "Banking and Interest Rates in a World without Money: The Effects of Uncontrolled Banking." *Journal of Bank Research* 1, 9-20.

Brito, Jerry and Andrea Castillo (2016), *Bitcoin: A Primer for Policymakers*, Fairfax, VA: Mercatus Center. [Online] http://mercatus.org/publication/bitcoin-primer-policymakers

Chapra, Mohammad Umer, "The Global Financial Crisis: Can Islamic Finance Help? " *Issues in the International Financial Crisis from an Islamic Perspective* , Jeddah: Islamic Economic Research Center (King Abdulaziz University), 11-18.

—— (1982), "Money and Banking in an Islamic Economy," in Mohammed Ariff, ed. *Monetary and Fiscal Economics of Islam*, Jeddah: Islamic Economic Research Center (King Abdulaziz University).

Choudhury, Masudul Alam and Md. Mostaque Hussain (2005), "A Paradigm of Islamic Money and Banking," *International Journal of Social Economics* 32(3), 203-217.

CoinDesk (2016-03-14), "Bitcoin Venture Capital," Website. http://www.coindesk.com/bitcoin-venture-capital/.

Commodity Futures Trading Commission (2014), "CFTC Issues Notice of Temporary Registration as a Swap Execution Facility to TeraExchange, LLC," *Release: PR6698-13*, [Online] http://www.cftc.gov/PressRoom/ PressReleases/pr6698-13

Cronin, David (2012), "The New Monetary Economics Revisited," *Cato Journal* 32(3), 581-594.

Del Mar, Alexander (1895), *History of Money*, London: Effingham Wilson, Royal Exchange. [Online] http://hdl.handle.net/2027/uc2.ark:/13960/t1fj2jd7r

El-Gamal, Mahmoud Amin (2006), "Overview of Islamic Finance," Occasional Paper No. 4, Office of International Affairs. [Online] http:// www.theiafm.org/publications/274_Islamic_Finance.pdf

European Central Bank (2012), *Virtual Currency Schemes*, Frankfurt: European Central Bank. [Online] http://www.ecb.europa.eu/pub/pdf/other/virtualcurrencyschemes201210en.pdf

Evans, Charles W. (2014a), "Virtual Monetary Economics," Working Paper [in submission].

——— (2014b), "Coins for Causes," Conscious Entrepreneurship Foundation White Paper (13 July). [Online] http://consciousentrepreneurship.org/coins-for-causes/

Fama, Eugene (1980) "Banking in the Theory of Finance," *Journal of Monetary Economics* 6(1), 39-57.

FinCEN (2013), "Application of FinCEN's Regulations to Persons Administering, Exchanging, or Using Virtual Currencies," *Guidance FIN-2013-G001*. [Online] http://www.fincen.gov/statutes_regs/guidance/html/FIN-2013-G001.html

Ghannadian, Farhad F., and Gautam Goswami (2004), "Developing Economy Banking: The Case of Islamic Banks," *International Journal of Social Economics* 31(8), 740-752.

Graef, Jean (2010), "IT Neologisms: Necessary but Dangerous," *Customizing and Governing the SharePoint Search System*. Available from the *Montague Institute Review* (January) https://web.archive.org/web/20141012043254/http://www.montague.co m:80/abstracts/neologisms.html

Greenfield, Robert L., and Leland B. Yeager (1983) "A Laissez-Faire Approach to Monetary Stability," *Journal of Money, Credit, and Banking* 15, 302–315.

Hall, Robert E. (1982a) "Explorations in the Gold Standard and Related Policies for Stabilizing the Dollar," *Inflation: Causes and Effects* (Robert E. Hall, Ed.), Chicago: University of Chicago Press, 111–122.

—— (1982b) "Monetary Trends in the United States and the United Kingdom: A Review from the Perspective of New Developments in Monetary Economics," *Journal of Economic Literature* 20, 1552–1556.

Hasan, Maher and Jemma Dridi (2010), "The Effects of the Global Crisis on Islamic and Conventional Banks: A Comparative Study," IMF Working Paper No. 10/201. [Online] http://www.imf.org/external/pubs/cat/lon-gres.aspx?sk=24183.0

Imam, Patrick and Kangni Kpodar (2013), "Islamic Banking: How Has It Expanded?" *Emerging Markets Finance & Trade* 49(6), 112–137.

International Organization for Standardization (2014), "Current Currency & Funds Code List," Zurich: Swiss Association for Standardization. [Online] http://www.currency-iso.org/en/home/tables/table-a1.html

Internal Revenue Service (2014), "IRS Virtual Currency Guidance : Virtual Currency Is Treated as Property for U.S. Federal Tax Purposes; General Rules for Property Transactions Apply," *Notice 2014-21*. [Online] http://www.irs.gov/pub/irs-drop/n-14-21.pdf

Khan, M. Mansoor and M. Ishaq Bhatti, (2008),"Islamic Banking and Finance: On Its Way to Globalization", *Managerial Finance* 34(10), 708-725.

Khan, Mohammad Fahim, "World Financial Crisis: Lesson form Islamic Economics ," *Issues in the International Financial Crisis from an Islamic Perspective* , Jeddah: Islamic Economic Research Center (King Abdulaziz University), 19-24.

Kirk, Jeremy (2013), "Could the Bitcoin Network Be Used as an Ultrasecure Notary Service?" PC World 24th May. [Online] http://www.pcworld.com/article/2039705/could-the-bitcoin-network-be-used-as-an-ultrasecure-notary-service.html

Larimer, Stan (2013), "Bitcoin and the Three Laws of Robotics," *Let's Talk Bitcoin* 14th September. [Online] http://letstalkbitcoin.com/bitcoin-and-the-three-laws-of-robotics/

Lavoie, Don & Emily Chamlee-Wright (2000), *Culture and Enterprise: The Development, Representation, and Morality of Business*, New York: Routledge.

Lo, Stephanie and J. Christina Wang (2014), "Bitcoin as Money?" Federal Reserve Bank of Boston *Current Policy Perspectives* 14-4, 1-28. [Online] http://www.bostonfed.org/economic/current-policy-perspectives/2014/cpp1404.htm

Matonis, Jon (2014), "12 Ways to Measure the Bitcoin Network's Health," *CoinDesk* (27th September). [Online] http://www.coindesk.com/12-ways-measure-bitcoin-networks-health/

Matthews, Robin and Issam Tlemsani (2010), "The Financial Roots of Babel: Roots of Crisis," *International Journal of Islamic and Middle Eastern Finance and Management* 3(4), 334-350.

Meikle, Scott (1994), "Aristotle on Money," *Phronesis* 39(1), 26-44.

Menger, Carl (1892), "On the Origin of Money," *Economic Journal* 2, 239–255.

Obaidullah, Mohammed (2005), *Islamic Financial Services*, Jeddah: Islamic Economic Research Center (King Abdulaziz University).

Owen, Olly (2009), "Biafran Pound Notes," *Africa* 79(4), 570-594.

Pluzhnyk, Angelina and Charles W. Evans (2014), "There's No Accounting for Bitcoin," Working Paper.

Radford, R.A. (1945), "The Economic Organization of a P.O.W. Camp," *Economica* 12, 189-201.

Rahn, Richard W. (1999), *The End of Money and the Struggle for Financial Privacy*, Seattle: Discovery Institute.

Rosenfeld, Meni (2012), "Overview of Colored Coins," Thesis. http://www.bibliothesis.com/thesis/2058001

Seidu, Abdullah Mohammed, "Current Global Financial Crisis: Cause and Solution ," *Issues in the International Financial Crisis from an Islamic Perspective* , Jeddah: Islamic Economic Research Center (King Abdulaziz University), 25-42.

Seyfang, Gill and Ruth Pearson (2000), "Time for Change: International Experience in Community Currencies," *Development* 43(4) , 56-60.

Siddiqui, Anjum (2008),"Financial Contracts, Risk and Performance of Islamic Banking", *Managerial Finance* 34(10), 680-694.

Siddiqi, Mohammad Nejatullah, "Current Financial Crisis And Islamic Economics ," *Issues in the International Financial Crisis from an Islamic Perspective* , Jeddah: Islamic Economic Research Center (King Abdulaziz University), 1-10.

—— (1988 [1969]), *Banking without Interest*, Leicester, England: Islamic Foundation.

Smolo, Edib and Abbas Mirakhor (2010), "The Global Financial Crisis and Its Implications for the Islamic Financial Industry," *International Journal of Islamic and Middle Eastern Finance and Management* 3(4), 372-385.

Taleb, Nassim Nicholas (2012), *Antifragile: Things That Gain from Disorder*, New York: Random House.

—— (2007), *The Black Swan: The Impact of the Highly Improbable*, New York: Random House.

—— (2005), *Fooled by Randomness*, New York: Random House.

Torpey, Kyle (2014), "Erik Voorhees Explains Why Some Bitcoin Companies Are Blocking American Users," *Let's Talk Bitcoin* 14th October. [Online] http://letstalkbitcoin.com/blog/post/erik-voorhees-explains-why-some-bitcoin-companies-are-blocking-american-users

Velde, François R. (2013), "Bitcoin: A Primer," Federal Reserve Bank of
   Chicago *Essays on Issues* 317, 1-4. [Online]
   http://www.chicagofed.org/webpages/publications/chicago_fed_letter/
   2013/december_317.cfm
Weber, Max (2001 [1905]), *The Protestant Ethic and the Spirit of
   Capitalism* , New York: Routledge Classics.
—— (1967 [1917-1919]), *Ancient Judaism*, New York: Free Press.
—— (2012 [1916]), *The Religion of India: The Sociology of Hinduism and
   Buddhism*, New Delhi: Munshiram Manoharlal.
—— (1968 [1915]), *The Religion of China: Confucianism and Taoism*, New
   York: Free Press.
White, Horace (1902 [1895]), *Money and Banking* 2nd ed., Boston: Ginn &
   Co.
World Bank (2012), "Three Quarters of the World's Poor Are 'Unbanked',"
   Data & Research (April 19th), Washington DC: World Bank. [Online]
   http://go.worldbank.org/72MAKHBAM0
Zarate, Juan C. (2013), *Treasury's War: Unleashing a New Era of Financial
   Warfare*, New York: PublicAffairs.

About the Author

Charles W. Evans taught Finance and Economics for 15 years and has been involved with virtual currencies since the first wave of *moneypunk* projects during the 1990s Dot-Com Era. In addition to advising entrepreneurs, he has worked over the past decade with plaintiff's attorneys in civil cases and prosecutors and defense attorneys in criminal cases related to money laundering, fraud, personal injury, and cryptocurrency.

A native of Miami, Dr. Evans lives once again in South Florida, and has lived in the New York City area, Metropolitan Washington DC, Los Angeles, southern Germany, West Berlin, Anguilla (Eastern Caribbean), and the Bahamas. He speaks German fluently and enough Spanish to get by.

Dr. Evans is available to clients worldwide and has been qualified as an expert on Bitcoin and cryptocurrency in several US State Courts, US Federal Court, and US Army Trial Judiciary. He can be reached at: cwe@EvansEconomics.com.